I0489278

NUREG–1449

Shutdown and Low-Power Operation at Commercial Nuclear Power Plants in the United States

Final Report

Date Published: September 1993

Office of Nuclear Reactor Regulation
U.S. Nuclear Regulatory Commission
Washington, DC 20555–0001

ABSTRACT

The report contains the results of the NRC staff's evaluation of shutdown and low-power operations at commercial nuclear power plants in the United States. The report describes studies conducted by the staff in the following areas: operating experience related to shutdown and low-power operations, probabilistic risk assessment of shutdown and low-power conditions, and utility programs for planning and conducting activities during periods the plant is shut down. The report also documents the staff's evaluations of a number of technical issues regarding shutdown and low-power operations, including the principal findings and conclusions. Potential new regulatory requirements are discussed, as are potential changes in NRC programs. This is a final report. Comments on the draft version of the report are discussed in Appendix C of this report.

CONTENTS

CONTENTS (cont.)

CONTENTS (cont.)

CONTENTS (cont.)

CONTENTS (cont.)

Appendices

Figures

Tables

EXECUTIVE SUMMARY

The NRC staff's evaluation of shutdown and low-power operations at commercial nuclear power plants in the United States is presented here. The study was initiated by the NRC investigation of the loss during shutdown of all vital ac power on March 20, 1990, at the Alvin W. Vogtle nuclear plant. This evaluation assessed risk broadly during shutdown, refueling, and startup, addressing not only issues raised by the Vogtle event, but also other issues related to shutdown that were identified by foreign regulatory organizations as well as by the NRC, and any new issues uncovered in the process.

The fundamental conclusion of this evaluation is that the public health and safety have been adequately protected while plants were in shutdown conditions, but that numerous and significant events have occurred indicating that substantial safety improvements are possible and appear warranted. The staff has also concluded, or perhaps reconfirmed, that reactor safety is the product of prudent, thoughtful, and vigilant efforts and not the result of "inherently safe" designs or "inherently safe" conditions. The areas of weakness identified herein stem primarily from the false premise that "shutdown" means "safe." The primary staff action resulting from this study must be, therefore, to recognize this fact and to resolve not to substitute complacency for appropriate safety programs to deal with shutdown conditions.

The evaluation was conducted in three stages. First, the NRC staff, with technical assistance from contractors, conducted technical studies to improve its understanding of the issues, and learned how the international community was dealing with the risks of shutdown.

Then, the staff integrated the findings from the technical studies to determine the most significant technical issues associated with shutdown, refueling, and startup operations, and to find topical areas that required further study. This process included a 3-day interoffice meeting of NRC personnel and their contractors to present issues and findings to date, followed by a peer assessment of the presentations conducted by the technical staff of the NRC Office of Nuclear Reactor Regulation (NRR).

In the third stage of the evaluation, the NRR technical staff responsible for the specific areas focused on assessing each of the key issues and study topics identified through the integration process. These assessments have yielded a number of potential regulatory actions to address the issues and the bases for those actions, as well as the bases for taking no action on some issues.

Throughout the course of the study, the NRC staff met periodically with the Nuclear Management and Re-

sources Council (NUMARC) to keep the industry informed of NRC activities and to keep NRC abreast of the industry's continuing initiatives. The staff met twice with the Advisory Committee on Reactor Safeguards (ACRS) to report its progress. The staff also briefed the Commission twice on the status of the evaluation and documented its progress in two Commission papers (SECY 91-283 and SECY 92-067).

On February 25, 1992, the staff issued this report in a draft form and requested public comment. The period for commenting ended on April 30, 1992. After the comment period ended, the staff held five public meetings to discuss the large number of comments received from utilities and industry organizations. At these meetings were representatives from each of the nuclear steam supply system (NSSS) owners groups, representatives from individual utilities, representatives from NUMARC, and members of the public. The staff has considered the public comments and, although the comments did not change any principal findings and conclusions, the staff did modify the report to clarify it and correct inaccuracies. In addition, the staff modified Chapter 7 of the draft report substantially based on the results of its ongoing regulatory analysis of shutdown issues. Staff responses to comments received on draft NUREG-1449 are provided in Appendix C.

The NRR had the major responsibility for conducting the evaluation. Other Headquarters offices, such as the Office of Nuclear Regulatory Research (RES), the Office for Analysis and Evaluation of Operational Data (AEOD), and regional offices gave strong support. Contractors assisting the staff were Brookhaven National Laboratory (BNL), Idaho National Engineering Laboratory (INEL), Science Applications International Corporation (SAIC), and Sandia National Laboratory (SNL).

Technical Studies

The NRC staff and its contractors completed the following studies as part of the evaluation:

- systematically reviewed operating experience, including reviewing reports of events at foreign and domestic operating reactors (AEOD)

- analyzed a spectrum of events at operating reactors to estimate the conditional probability of core damage using the accident sequence precursor (ASP) analysis methodology (SAIC for NRR)

- visited 11 plant sites to broaden staff understanding of shutdown operations, including outage planning, outage management, and startup and shutdown activities (NRR)

- reviewed and evaluated existing domestic and foreign probabilistic risk assessments (PRAs) that address shutdown conditions (NRR)

- completed a preliminary level 1 PRA of shutdown and low-power operating modes for a pressurized-water reactor (PWR) and a boiling-water reactor (BWR) to screen for important accident sequences (BNL and SNL for RES)

- completed thermal-hydraulic scoping analyses to estimate the consequences of an extended loss of residual heat removal (RHR) in PWRs, and evaluated alternate methods of RHR (INEL for NRR)

- completed an analysis to estimate the likelihood and consequences of a rapid, non-homogeneous dilution of borated water in a PWR reactor core (BNL for NRR)

- compiled and reviewed existing regulatory requirements for shutdown operation and important safety-related equipment (SAIC for NRR)

- met with specialists from the Organization for Economic Cooperation and Development/Nuclear Energy Agency to exchange information on current regulatory approaches to the shutdown issues in member countries and drafted a paper on the various approaches (NRR)

The details and findings of these studies are discussed in Chapters 2, 3, 4, 5, and 6 of the report.

The most significant technical findings from the evaluation are the following:

- Outage planning is crucial to safety during shutdown conditions since it establishes if and when a licensee will enter circumstances likely to challenge safety functions, and the level of mitigation equipment available.

- The current NRC requirements in the area of fire protection (i.e., 10 CFR Part 50, Appendix R) do not apply to shutdown conditions. However, significant maintenance activities, which can increase the potential for fire, do occur during shutdown.

- Well-trained and well-equipped plant operators can play a very significant role in accident mitigation for shutdown events.

- All probabilistic risk assessments for shutdown conditions in PWRs find that accident sequences involving loss of RHR during operation with a reduced inventory (e.g., midloop operation) are dominant contributors to the core-damage frequency.

- Extended loss of decay heat removal capability in PWRs can lead to a loss-of-coolant accident (LOCA) caused by failure of temporary pressure boundaries in the reactor coolant system (RCS) or rupture of RHR system piping. In either case, the containment may be open and emergency core cooling system (ECCS) recirculation capability may not be available.

- Passive methods of decay heat removal can be very effective in delaying or preventing a severe accident in a PWR; however, there are no procedures or training for such methods.

- All PWR and Mark III BWR primary containments are capable of providing significant protection under severe core-damage conditions, provided that the containment is closed or can be closed quickly. However, analyses have shown that the steam and radiation environment in containment, which can result from an extended loss of RHR or LOCA, would make it difficult to close the containment. Mark I and II BWR secondary containments offer little protection, but this is offset by a significantly lower likelihood of core damage in BWRs.

- Generation of a dilute water slug in the RCS of a PWR during startup is possible but very unlikely. The effect of such a slug moving through the core would be limited to a power excursion which could result in some fuel damage but not a breach of the reactor vessel.

Potential Industry Actions Being Evaluated With Regulatory Analysis

In the draft version of NUREG–1449 issued for comment in early 1992, the staff identified the following five areas in which improvements in shutdown operations appeared to be warranted:

(1) outage planning and control

(2) fire protection

(3) operations, training, procedures, and other contingency plans

(4) technical specifications

(5) instrumentation

Since issuing that draft, the staff has been conducting a formal regulatory analysis to determine which, if any, improvements could be justified as backfits. In conducting its formal regulatory analysis, the staff performed qualitative as well as quantitative evaluations of the five items above as a combined comprehensive program for conducting shutdown activities at either PWRs or BWRs. Such a program would be governed by two main catego-

controls, which include (1) administrative controls for activities related to organization, management, and procedures and (2) limiting conditions for operation (LCOs) for controlling the availability of equipment needed to mitigate an accident. The staff views programmatic and procedural actions related to items 2, 3, and 5 (listed above) as administrative controls incorporated into the process of planning and controlling outages. Item 4 above, i.e., technical specifications, would include only LCOs on equipment. Specific controls being evaluated by the staff are listed below and in Table 1. In addition to the technical specifications controls, the staff is evaluating a hardware improvement to enhance the capability for monitoring the reactor vessel water level during operation with a reduced inventory.

Improvements in Planning and Controlling Outages (PWR and BWR)

Licensees can improve their programs for planning and controlling outages by incorporating new and improved administrative controls and LCOs. Licensees may be required to develop and use a program for planning and controlling outages that would include those elements listed below. In addition, licensees may be required to adopt new technical specifications with LCOs similar to those in Table 1.

The staff considers the programmatic guidelines in NUMARC 91-06 to address those elements of an outage program listed below with the notable exceptions being element 7 (instrumentation) and element 9 (specific contingency plans for fire protection). Consequently, the staff believes that a licensee program that (1) fully implements the guidelines in NUMARC 91-06 and (2) incorporates the features regarding fire protection and instrumentation listed below would be consistent with the staff's assumptions regarding the administrative controls portion of this improvement.

Elements for an Outage Program

(1) clearly defined and documented safety principles for outage planning and control

(2) clearly defined organizational roles and responsibilities

(3) controlled procedure defining the outage planning process

(4) pre-planning for all outages

(5) strong technical input based on safety analysis, risk insights, and defense in depth

(6) independent safety review of the outage plan and subsequent modifications

(7) planning and controls that (a) maximize the availability of existing instrumentation used to monitor temperature, pressure, and water level in the reactor vessel and (b) provide accurate guidelines for operations when existing temperature indications may not accurately represent core conditions

(8) controlled information system to provide critical safety parameters and equipment status on a real-time basis during the outage

(9) contingency plans and bases, including those necessary to ensure that effective decay heat removal (DHR) during cold shutdown and refueling conditions can be maintained in the event of a fire in any plant area

(10) realistic consideration of staffing needs and personnel capabilities with emphasis on control room staff

(11) training

(12) feedback of shutdown experience into the planning process

Improvements in PWR Instrumentation

Licensees of PWRs may be required to have an independent, diverse means of accurately monitoring reactor vessel water level during midloop operation that provides continuous indication in the control room and an alarm to alert operators to over-draining during an approach to a midloop condition (e.g., ultrasonic or local pressure differential measurements across the hot leg).

Staff Actions

During the course of the evaluation, the staff has taken a number of actions in response to concerns about shutdown operations. These actions include issuing information notices regarding shutdown operations, use of freeze seals, and the potential for boron dilution. In addition, the staff issued a temporary instruction (TI) calling for increased inspection emphasis during outages that focused primarily on RHR capability and activities involving electrical systems. To fully develop the TI, the staff has conducted pilot inspections at Oconee Unit 2, Indian Point Unit 3, Diablo Canyon Unit 1, Prairie Island Units 1 and 2, and Cooper station. The staff has also modified NRC standards for operator license exams to (1) place more emphasis on shutdown operations and (2) review the licensee's requalification exam test outline for coverage of shutdown and low-power operations, consistent with the licensee's job task analysis and operating procedures. Finally, Headquarters staff advised regional staff that current emergency plans should address protection of plant

Table 1 Limiting Conditions for Operation During Cold Shutdown and Refueling

System	PWR LCOs		BWR LCOs	
	Mode 5 Mode 6 Low Level	Mode 6 High Level	Mode 4 Mode 5 Low Level	Mode 5 High Level
Residual heat removal	2 trains OPERABLE*	1 train OPERABLE*	2 trains OPERABLE*	1 train OPERABLE*
Emergency core cooling	2 trains OPERABLE	Not required*	2 trains OPERABLE*	Not required*
Offsite ac power	1 offsite source OPERABLE*	1 offsite source OPERABLE*	1 offsite source OPERABLE*	1 offsite source OPERABLE*
Onsite ac Power	2 onsite sources OPERABLE	1 onsite source OPERABLE*	2 onsite sources OPERABLE	1 onsite source OPERABLE*
Primary containment integrity	Required when decay heat rate is > [ ] and RCS temperature is > [ ]	Not required*	Not required*	Not required*
Service water	2 trains OPERABLE	1 train OPERABLE	2 trains OPERABLE	1 train OPERABLE
Equipment cooling water	2 trains OPERABLE	1 train OPERABLE	2 trains OPERABLE	1 train OPERABLE

*Currently specified in Standard Technical Specifications.

workers in an emergency during shutdown operations. The staff has also identified a number of potential actions that are discussed in Chapter 8 of the report. They include

- Incorporate findings from shutdown and low-power evaluation into licensing reviews for advanced light-water reactors.

- Continue level 1 and level 2 PRA studies of shutdown and low-power operations at Grand Gulf and Surry.

- Continue evaluation of pilot team inspections for shutdown operations and report findings and recommendations to the Commission.

- Develop a performance indicator for shutdown operations to monitor licensee performance in this area and incorporate the results in NUREG-1022.

- Develop and issue interim guidance for classifying accidents that occur during shutdown.

The staff has identified a number of safety issues important to shutdown and low-power operation. Resolving these issues through new generic requirements could improve safety substantially. The staff bases this conclusion on observations and inspections at a number of plants, deterministic safety analysis, insights gained from probabilistic risk assessments, and some quantitative risk assessment. In accordance with the shutdown-risk program plan and schedule, the staff is continuing to assess the need for regulatory action on low-power and shutdown issues, including analyses in accordance with the backfit rule, 10 CFR 50.109.

1 BACKGROUND AND INTRODUCTION

Over the past several years, the Nuclear Regulatory Commission (NRC) staff has become more concerned about the safety of operations during shutdown. The Diablo Canyon event of April 10, 1987, highlighted the fact that the operation of a pressurized-water reactor (PWR) with a reduced inventory in the reactor coolant system presented a particularly sensitive condition. From NRC's review of the event, the staff issued Generic Letter 88-17 on October 17, 1988. The letter requested that licensees address numerous generic deficiencies to improve safety during operation at reduced inventory. More recently, the incident investigation team's report of the loss of ac power at the Vogtle plant (NUREG-1410) emphasized the need for risk management of shutdown operations. Furthermore, discussions with foreign regulatory organizations (i.e., French and Swedish authorities) about their evaluations regarding shutdown risk have reinforced previous NRC staff findings that the core-damage frequency for shutdown operation can be a fairly substantial fraction of the total core-damage frequency. Because of these concerns regarding operational safety during shutdown, the staff began a careful, detailed evaluation of safety during shutdown and low-power operations.

On July 12, 1990, the staff briefed the Advisory Committee on Reactor Safeguards (ACRS) on its draft plan for a broad evaluation of risks during shutdown and low-power operation. On October 22, 1990, the staff issued the plan in the form of a memorandum from James M. Taylor, to the Commissioners, "Staff Plan for Evaluating Safety Risks During Shutdown and Low Power Operations." The staff briefed the ACRS on the status of the evaluation on June 5 and 6, 1991, and on June 19, 1991, the staff discussed the status of the evaluation in a public meeting with the Commission. On September 9, 1991, the staff issued a Commission paper (SECY 91-283) which reported progress to date on the evaluation and provided a detailed plan for addressing each of the technical issues identified.

1.1 Scope of the Staff Evaluation

In the staff's evaluation, "shutdown and low-power operation" encompasses operation when the reactor is in a subcritical state or is in transition between subcriticality and power operation up to 5 percent of rated power. The evaluation addresses only conditions for which there is fuel in the reactor vessel (RV). The evaluation addresses all aspects of the nuclear steam supply system (NSSS), the containment, and all systems that support operation of the NSSS and containment. However, the evaluation does not address events involving fuel handling outside of the containment, fuel storage in the fuel storage building, and events that do not involve the previously identified systems.

1.2 Organization

The Office of Nuclear Reactor Regulation (NRR) has the lead responsibility for conducting the evaluation. However, other Headquarters offices, such as the Office of Nuclear Regulatory Research (RES), the Office for Analysis and Evaluation of Operational Data (AEOD), and regional offices have contributed strong support. A group of senior managers representing these offices served as the steering committee for the evaluation. This group met periodically to be briefed on the progress of the evaluation and to provide guidance. Members of the steering committee included the following: William Russell, Associate Director for Inspection and Technical Assessment, NRR; Ashok Thadani, Director, Division of Systems Technology, NRR; Brian Sheron, Director, Division of Systems Research, RES (later replaced by Warren Minners, Director, Division of Safety Issue Resolution); Samuel Collins, Director, Division of Reactor Projects, Region IV; and Thomas Novak, Director, Division of Safety Programs, AEOD.

1.3 Summary of the Evaluation

In its original plan, the staff divided work necessary to complete the evaluation into six major elements containing a number of interrelated tasks to be completed over 18 months. The six major program elements are the following:

I. Review and evaluate event experience and event studies.

II. Study shutdown operations and activities.

III. Conduct probabilistic risk assessment (PRA) activities and engineering studies.

IV. Integrate technical results to understand risk.

V. Evaluate guidance and requirements affecting risk management.

VI. Recommend new regulatory requirements as necessary.

Consistent with this program plan, the staff and its contractors have completed the following studies which, as indicated, are fully discussed later in this report:

• systematically reviewed operating experience, including reviewing reports of events at foreign and domestic operating reactors, and documented the findings in the AEOD engineering evaluation (Chapter 2)

- with assistance from the Science Applications International Corporation (SAIC), analyzed a spectrum of events at operating reactors using the accident sequence precursor methodology (Chapter 2)

- visited 11 plant sites to broaden staff understanding of shutdown operations, including outage planning, outage management, and startup and shutdown activities (Chapter 3)

- reviewed, evaluated, and documented the few existing domestic and foreign PRAs that address shutdown conditions (Chapter 4)

- completed and documented a coarse level 1 PRA of shutdown and low-power operating modes for a PWR and a boiling-water reactor (BWR) through RES contractors at Brookhaven National Laboratory and Sandia National Laboratory (Chapter 4)

- with technical assistance from the Idaho National Engineering Laboratory, completed and documented several thermal-hydraulic studies that address the consequences of an extended loss of residual heat removal (Chapter 6)

- with assistance from Brookhaven National Laboratory, completed and documented an analysis to estimate the likelihood and consequences of a rapid non-homogeneous dilution of borated water in a PWR reactor core (Chapter 6)

- with technical assistance from SAIC, compiled existing regulatory requirements for shutdown operation and important safety-related equipment (Chapter 5)

- coordinated a meeting with specialists from the Organization for Economic Cooperation and Development/Nuclear Energy Agency to exchange information on current regulatory approaches to the shutdown issues in member countries, including drafting a discussion paper on the various approaches (Chapter 5)

- met periodically with the Nuclear Management and Resources Council to keep the industry informed of NRC activities and to stay abreast of the industry's continuing initiatives

To integrate its findings from these studies and to define important technical issues, the staff met for three days with contractors from several national laboratories who had been working on the shutdown and low-power evaluation or had special expertise in the issue. During this meeting, held April 30 through May 2, 1991, the staff identified five issues that are especially important for shutdown and a number of additional topics that warrant further evaluation. These issues are

- outage planning and control

- stress on personnel and programs

- training and procedures

- technical specifications

- PWR safety during midloop operation

Topics identified for further evaluation included the following:

- loss of residual heat removal capability

- containment capability

- rapid boron dilution

- fire protection

- instrumentation

- emergency core cooling system recirculation capability

- effect of PWR upper internals

- onsite emergency planning

- fuel handling and heavy loads

- potential for draining the BWR reactor vessel

- reporting requirements for shutdown events

- need to strengthen inspection program

The staff proposed an evaluation plan for each of the issues and topics and documented the plans in a Commission paper issued September 9, 1991 (SECY 91-283). The evaluations are now complete and the results form the basis for the staff's technical findings and conclusions given in Chapter 6, and recommended actions given in Chapters 6, 7, and 8 of this report. However, it should be noted that Chapters 7 and 8 have been revised substantially from the earlier draft version of the report issued for comment in February 1992. Comments on the draft version are listed and discussed in Appendix C.

2 ASSESSMENT OF OPERATING EXPERIENCE

2.1 Retrospective Review of Events at Operating Reactors

The staff reviewed operating experience to ensure that its evaluation encompassed the range of events encountered during shutdown and low-power operation: licensee event reports (LERs), studies performed by the Office for Analysis and Evaluation of Operational Data (AEOD), and various inspection reports to determine the types of events that take place during refueling, cold and hot shutdown, and low-power operation.

The staff also reviewed events at foreign nuclear power plants using information found in the foreign events file maintained for AEOD at the Oak Ridge National Laboratory. The AEOD compilation included the types of events that applied to U.S. nuclear plants and those not found in a review of U.S. experience.

In performing this review, the staff found that the more significant events for pressurized-water reactors (PWRs) were the loss of residual heat removal, potential pressurization, and boron dilution events. The more important events for boiling-water reactors (BWRs) were the loss of coolant, the loss of cooling, and potential pressurization. Generally, the majority of important events involved human error—administrative, other personnel, or procedural. In December 1990, the staff documented this review in the AEOD special report, "Review of Operating Events Occurring During Hot and Cold Shutdown and Refueling," which is summarized below. In addition, the staff selected 10 events from the AEOD review for further assessment as precursors to potential severe core-damage accidents. This assessment is discussed in Section 2.2.

The AEOD special report encompassed events that had occurred primarily between January 1, 1988, and July 1, 1990. An initial database was created which included 348 events gathered primarily from the Sequence Coding and Search System and significant events that occurred before or after the target period. Of the 348 events, approximately 30 percent were considered more significant and were explicitly discussed in the AEOD report.

The events were evaluated by plant type (i.e., PWR or BWR) and six major event categories: loss of shutdown cooling, loss of electrical power, containment integrity problems, loss of reactor coolant, flooding and spills, and overpressurization of the reactor coolant system; for PWRs, boron problems were also included. Less frequently occurring events, such as fires, were covered briefly.

The results of the AEOD study are discussed in Sections 2.1.1 through 2.1.7. Insights gained from the study are given in Section 2.1.8.

2.1.1 Loss of Shutdown Cooling

The loss of shutdown cooling is one of the more serious event types and can be initiated by the loss of flow in the residual heat removal (RHR) system or by loss of an intermediate or ultimate heat sink. Events involving loss of cooling that occur shortly after plant shutdown may quickly lead to bulk boiling and eventual fuel uncovery if cooling is not restored.

The evaluation included 16 PWR and 11 BWR events involving loss of shutdown cooling; these are listed in Tables 2.1 and 2.2.

More than 60 percent of the PWR events arose from human error—administrative, other personnel, or procedural. Equipment problems accounted for 16 percent of the events. The types of incidents that caused the events ranged from the RHR pump becoming air bound, through loss of power to the RHR pump, to the malfunction of level indication in the control room. These events resulted in temperature rises ranging from 15° to 190° (on the Fahrenheit scale) (–9.4° to 88° on the Celsius scale).

For the BWR events, approximately 60 percent were caused by human error—administrative, other personnel, or procedural.

Table 2.1 Events Involving PWR Loss of Shutdown Cooling

Plant	Event date
Millstone 2	12/09/81
Salem 1	03/16/82
Catawba 1	04/22/85
Zion 2	12/14/85
Crystal River 3	02/02/86
Waterford 3	07/14/86
Diablo Canyon 2	04/10/87
Oconee 3	12/16/87
Oconee 3	09/11/88
Arkansas 1	10/26/88
McGuire 1	11/23/88
Arkansas 1	12/19/88
Braidwood 2	01/23/89
Salem 1	05/20/89
Arkansas 1	12/06/89
Vogtle 1	03/20/90

Table 2.2 Events Involving BWR Loss of Shutdown Cooling	
Plant	Event date
Brunswick 1&2	04/17/81
Susquehanna 1	03/21/84
Fermi 2	03/18/88
FitzPatrick	10/21/88
Susquehanna 1	01/07/89
River Bend	06/13/89
Pilgrim	12/09/89
Duane Arnold	01/09/90
FitzPatrick	01/20/90
Susequehanna 1	02/03/90

Table 2.3 Events Involving PWR Loss of Reactor Coolant	
Plant	Event date
Haddam Neck	08/21/84
Farley 2	10/27/87
Surry 1	05/17/88
Sequoyah 1	05/23/88
San Onofre 2	06/22/88
Byron 1	09/19/88
Cook 2	02/16/89
Indian Point 2	03/25/89
Palisades	11/21/89
Braidwood 1	12/01/89

2.1.2 Loss of Reactor Coolant Inventory

The chance that reactor coolant will be lost from the reactor vessel can actually increase during shutdown modes because large, low-pressure systems, such as RHR, are connected to the reactor coolant system. The safety significance of such loss is that it could lead to voiding in the core and eventual core damage.

The evaluation included 22 events involving loss of reactor coolant. The plants and dates of the events are listed in Tables 2.3 and 2.4.

The PWR events had various causes, such as opening of the RHR pump suction relief valve, power-operated relief valve (PORV) and block valves opening simultaneously during PORV testing, and loss of pressure in the reactor cavity seal ring allowing drainage from the cavity. These events accounted for losses of reactor coolant inventory of up to 67,000 gallons (254 kL).

Many of the BWR events included in the evaluation were caused by valve lineup errors and resulted in decreased levels of up to 72 inches (183 cm).

Of the 10 PWR events reported in the AEOD evaluation, 6 were caused by human errors and 4 were caused by equipment problems. Of the 12 BWR events included in the evaluation, 10 were caused by human errors and only 2 were caused by equipment failure.

2.1.3 Breach of Containment Integrity

A breach of containment integrity in itself may not be of great safety significance, but this condition, coupled with postulated events, could substantially increase the severity of the event. Also, a breach of containment integrity in conjunction with fuel failure could cause the release of radioactive material. Eight events involving breach of containment were included in the AEOD evaluation. All were due to human error.

2.1.4 Loss of Electrical Power

The safety significance of the loss of electrical power depends on the part of the plant affected. The loss could range from complete loss of all ac power to the loss of a dc bus or an instrument bus. Loss of electrical power generally leads to other events, such as loss of shutdown cooling.

The events included in the AEOD evaluation are listed in Table 2.5.

Of the 13 PWR events evaluated by AEOD, 7 were caused by human errors, 5 were caused by maintenance, and 1 was caused by fire. Of the original 45 events found in the AEOD study, approximately 62 percent were caused by human error and approximately 20 percent were caused by equipment problems. The BWR statistics were reversed: only 20 percent of the events were caused by human errors and 50 percent were caused by equipment problems.

Table 2.4 Events Involving BWR Loss of Reactor Coolant	
Plant	Event date
Grand Gulf	04/03/83
LaSalle 1	09/14/83
LaSalle 2	03/08/84
Washington Nuc 2	08/23/84
Susquehanna 2	04/27/85
Hatch 2	05/10/85
Peach Bottom 2	09/24/85
Fermi 2	03/13/87
Washington Nuc 2	05/01/88
Pilgrim	12/03/88
Vermont Yankee	03/09/89
Limerick	04/07/89

Table 2.5 Events Involving Loss of Electrical Power

PWR	Event date	Description of event
Turkey Point 3	05/77/85	Loss of offsite power
Fort Calhoun	03/21/87	Loss of all ac offsite power
McGuire 1	09/16/87	Loss of offsite power
Harris	10/11/87	Loss of power to safety buses
Wolf Creek	10/15/87	Loss of 125-V dc source
Crystal River 3	10/16/87	Loss of power to one of two vital buses
Indian Point 2	11/05/87	Loss of power to the 480-V ac bus
Braidwood 2	01/31/88	Instrument bus deenergized
Millstone 2	02/04/88	Loss of power to vital 4160-V ac train
Yankee Rowe	11/16/88	Loss of power to two emergency 480-V buses
Oconee 3	09/11/88	Loss of ac power to shutdown cooling equipment
Fort Calhoun	02/26/90	Loss of power to 4160-V safety buses
Vogtle 1	03/20/90	Loss of offsite and onsite ac power sources

BWR		
Pilgrim	11/12/87	Loss of offsite power
Nine Mile 2	12/26/88	Loss of offsite power
Millstone 1	04/29/89	Loss of normal power
Washington Nuclear 2	05/14/89	Loss of offsite power
River Bend	03/25/89	Division II loss of power
Limerick	03/30/90	Loss of a power supply

2.1.5 Overpressurization of Reactor Coolant System

Both PWR and BWR overpressurization events have occurred during shutdown conditions. Such events are precursors to exceeding the reactor vessel brittle fracture limits or the American Society of Mechanical Engineers Boiler and Pressure Vessel Code (ASME Code) limits. The reactor coolant system (RCS) generally overpressurizes in one of three ways: operation with the RCS completely full and experiencing pressure control problems, occurrences of inadvertent safety injection, or pressurization of systems attached to the RCS.

Of the significant events considered in the AEOD evaluation, there were not enough to indicate a trend regarding the cause of the events. However, the original database included 24 PWR pressurization events, and 66 percent of those events had been caused by human errors. Only three BWR events were in the original database.

2.1.6 Flooding and Spills

The safety significance of flooding or spills depends on the equipment affected by the spills. The AEOD evaluation included 3 of the 29 PWR events in the original database. Of the original 29 PWR events, more than 50 percent were caused by human errors; 14 percent were caused by equipment problems. There were only 7 BWR flooding or spill events in the original database and the majority were caused by human errors.

2.1.7 Inadvertent Reactivity Addition

Both PWR and BWR plants had experienced inadvertent criticalities, some of which resulted in reactor scrams. The AEOD evaluation indicated that inadvertent reactivity addition in PWRs was caused primarily by dilution while the plant was shut down. Also boron dilution without the operator's knowledge was identified as a potentially severe event. In BWRs, inadvertent reactivity addition was most often caused by human error (the operator selected the wrong control) and feedwater transients.

The events included in the evaluation are listed in Table 2.6.

2.1.8 Insights From the Review of Events

The original database of shutdown events included 348 events, most of which had occurred since 1985. AEOD used experience and engineering judgment in selecting those that were the more significant. Those 30 significant events were then categorized to help AEOD determine the cause and identify any trending.

Two major observations became apparent in the evaluation whether using the original database of 348 or the narrowed database of 30 more significant events. The first observation is that a greater percentage of the events were caused by human errors than by equipment problems. The second observation is that the events did not reveal new unanalyzed issues but, instead, appeared to represent an accumulation of errors or equipment failures or a combination of the two.

Table 2.6 Events Involving Inadvertent Reactivity Addition

PWR	Event date	Description of event
Surry 2	04/14–23/89	Boron concentration decreased by leak in RCP stand pipe makeup valve
Turkey Point 3&4	05/28–06/03/87	Unable to borate Unit 3 volume control tank (VCT) because of nitrogen gas binding of all boric acid transfer pumps
Arkansas 2	05/04/88	Gas binding of the charging pumps from inadvertent emptying of the VCT
Foreign reactor	1990	Boron dilution from a cut steam generator tube thathad not been plugged

BWR		
Millstone 1	11/12/76	Withdrawal of the wrong control rod and a suspected high worth rod
Browns Ferry 2	02/22/84	Withdrawal of high worth rod
Hatch 2	11/7/85	Feedwater transient
Peach Bottom 3	03/18/86	Incorrect rod withdrawn
River Bend	07/14/86	Feedwater transient
Oyster Creek	12/24/86	Feedwater transient

2.2 Accident Sequence Precursor Analysis

Using the accident sequence precursor (ASP) method, the staff and its contractors, Oak Ridge National Laboratory and Science Applications International Corporation, evaluated a sample of 10 shutdown events that could be significant. The staff reviewed this sample to determine the conditional probability of core damage, that is, the probability of core damage, given that the initiating event has already occurred, from each type of event selected in order to help characterize the overall shutdown risk for U.S. nuclear power plants. As discussed in Section 2.2.1, the 10 selected events reasonably represented the reactor population of BWRs, PWRs, and the various vendors.

To date, the ASP program has been largely concerned with operational events that occurred at power or hot shutdown. Methods used in that program to identify operational events considered precursors, plus the models used to estimate risk significance, have been developed over a number of years. In particular, the ASP core-damage models have been improved over time to reflect insights from a variety of probabilistic risk assessment studies. In applying ASP methods to evaluate events during cold shutdown and refueling, the same analytical approach was used. However, accident sequence models describing failure combinations leading to core damage had to be developed, with little earlier work as a basis.

This analysis was exploratory in nature. Its intent was to ensure that operating experience was assessed systemati-cally (1) to develop insights into (a) the types of events that have occurred during shutdown and (b) which characteristics of these events are important to risk, and (2) to develop methods that could be used in a continuing manner to analyze shutdown events. The staff did not intend to use this effort to make comparisons with analyses of at-power events in the ASP program.

The following section describes how the 10 events that were analyzed were selected. Section 2.2.2 summarizes the development of core-damage models and the estimation of conditional probabilities. Finally, Section 2.2.3 describes the results of the analyses and overall findings. The complete detailed analysis for each event is documented in Appendix A.

2.2.1 Selecting Events for Analysis

The staff selected 10 events that had occurred during cold shutdown and refueling for analysis. The staff chose these events after it had (1) reviewed the AEOD evaluation of non-power events discussed in Section 2.1 and (2) performed confirmatory searches using the Sequence Coding and Search System, a database of LER information maintained at Oak Ridge National Laboratory (ORNL).

Events chosen were considered representative of the types of events that could impact shutdown risk and that could be analyzed using ASP methods. These events concerned loss of reactor inventory, loss of residual heat removal, and loss of electric power. One event involved a flood that had safety system impacts. The events chosen for analysis were considered potentially more serious than the typical event observed at cold shutdown.

Events were also chosen so that all four reactor vendors were represented in the analysis. This allowed the staff to explore modeling issues unique to different plant designs and to develop models that could be applied at a later date to a broad set of cold-shutdown and refueling events.

The 10 events chosen for analysis are listed in Table 2.7. The 10 events are sorted by date and by vendor in Table 2.8. The 1990 loss of ac power and shutdown cooling (SDC) at Vogtle 1 is not included in the list because it was evaluated previously with the ASP methodology as discussed in NUREG/CR-4674.

2.2.2 Analysis Approach

The staff analyzed each of the events listed in Table 2.7. This analysis included a review of available information concerning each event and plant to determine system lineups, equipment out of service, water levels and reactor pressure vessel (RPV) inventories, time to boil and to core uncovery, vessel status, and so on. This involved review of final safety analysis reports, augmented inspection team reports, operating procedures, and supplemental material in order to understand the system interactions that occurred during the event, the recovery actions and alternate strategies that could be employed, and the procedures available to the operators.

Once the event had been characterized and its effect on the plant was understood, event significance was estimated based on methods used in the ASP program. Quantification of event significance involves determining a conditional probability of subsequent core damage given the failures that occurred. (See Section 2.2.3 for the current limitations in this approach.) The conditional probability estimated for each event is important because conditional probability provides an estimate of the measure of protection remaining against core damage once the observed failures have taken place. Conditional probabilities were estimated by mapping failures observed during the event onto event trees that depict potential paths to severe core damage, and by calculating a conditional probability of core damage through the use of event tree branch probabilities modified to reflect the event. The effect of an event on event tree branches was assessed by reviewing the operational event specifics against system design information and translating the results of the review into a revised conditional probability of branch failure given the operational event.

In the ASP analysis, only sequential events that can occur after the failures that actually occurred in the accident are modeled. Consequently, in the quantification process, "failure" probabilities, i.e., those in the downward direc-

tion, in the event trees for systems observed to have failed during the actual accident reflect only the likelihood of not recovering from the failure or fault that actually occurred. Failure probabilities for systems observed to have degraded during the actual operational event were assumed equal to the conditional probability that the system would fail (given that it was observed degraded) and the probability that it would not be recovered within the required time period. The failure probabilities associated with observed successes and with systems unchallenged during the actual event were assumed equal to a failure probability estimated by the use of system success criteria and train and common-mode failure screening probabilities, with consideration of the potential for recovery.

Event tree models were developed to describe potential core-damage sequences associated with each event. For the purposes of simplifying this analysis, core damage was conservatively assumed to occur when RPV water level decreased to below the top of active fuel. Choice of this damage criterion allowed the use of simplified calculations to estimate the time to an unacceptable end state. Core damage was also assumed to occur if a combination of systems, as specified on the event tree, failed to perform at a minimum acceptable level and could not be recovered.

The event tree model used to analyze an event was developed on the basis of procedures that existed then. These procedures were considered the primary source of information available to the operators concerning the steps to be taken to recover from the event or to implement another strategy for cooling the core. Since procedures varied greatly among plants, the event trees developed to quantify an event were typically plant and event specific. Event trees applicable to each analysis are described in Appendix A.

In developing branch probability estimates for the cold-shutdown models, the probability of not recovering a faulted branch before boiling or core uncovery occurred frequently had to be estimated. Applicable time periods were often 6 to 24 hours.

There are no operator response models (especially models out of the control room) or equipment repair models for these time periods. For the purposes of this analysis, the probability of crew failure as a function of time for non-proceduralized actions was developed by skewing applicable curves for knowledge-based action in the control room by 20 minutes to account for recovery time outside the control room. A minimum (truncated) failure probability of 1×10^{-4} was also specified. For long-term proceduralized actions, recovery was assumed to be dominated by equipment failure, and operator failure was not addressed. The probability of failing to repair a faulted system before boiling or core uncovery occurred was estimated using an exponential repair model with the observed repair time as the median.

Table 2.7 Cold-Shutdown and Refueling Events Analyzed Using ASP Methods, by Docket/LER No.

Docket/ LER No.	Description of event (date)	Conditional core-damage probability*
271/89–013	10,000 gal of reactor vessel inventory was transferred to the torus at Vermont Yankee when maintenance stroked-tested the SDC valves in the but-of-service loop of RHR with the minimum flow valve already open. More than 45 min required to locate and isolate the leak. (3/9/89)	1×10^{-6}
85/90–006	Loss of offsite power with the emergency diesel generators not immediately available at Fort Calhoun. Breaker failure relay operated to strip loads, but EDG design feature prevented auto loading. (2/26/90)	4×10^{-4}
287/88–005	Loss of ac power and loss of RHR during midloop operation with vessel head on at Oconee 3. Testing errors caused a loss of power to feeder buses resulting in loss of SDC with no accompanying reactor temperature or level indication. (9/11/88)	2×10^{-6}
302/86–003	RHR pump shaft broke during midloop operation at Crystal River 3. Pump had been in continuous operation for about 30 days. A tripped circuit breaker delayed placing the second train on line. (2/2/86)	1×10^{-6}
323/87–005	Loss of RHR at Diablo Canyon 2 while at midloop operation. RCS inventory lost through a leaking valve and air entrainment in both RHR pumps caused loss of SDC. Extended boiling occurred. (4/10/87)	5×10^{-5}
382/86–015	Loss of RHR during midloop operation at Waterford 3. Complications in restoring RHR due to steam binding and RHR pump suction line design. Extended boiling occurred. (7/14/86)	2×10^{-4}
387/90–005	Extended loss of RHR at Susquehanna 1. An electrical fault caused isolation of SDC suction supply to RHR system. Alternate RHR provided using the suppression pool. (2/3/90)	3×10^{-5}
397/88–011	Loss of reactor vessel inventory at Washington Nuclear Plant 2 (WNP–2). The RHR suppression pool suction and SDC suction valves were open simultaneously, and approximately 10,000 gal of reactor water was transferred to the suppression pool. (5/1/88)	5×10^{-5}
456/89–016	RCS inventory loss at Braidwood 1. An RHR suction relief valve stuck open and drained approximately 64,000 gal of water from the RCS before being isolated. (12/1/89)	1×10^{-6}
458/89–020	15,000 gal (57.8 kL) of service water flooded the auxiliary building when a freeze seal failed at River Bend. One RHR train, normal spent fuel pool cooling, and auxiliary and reactor building building lighting lost. (4/19/89)	1×10^{-6}

*See Section 2.2.3 for the limitations to this approach.

Table 2.8 Cold-Shutdown and Refueling Events Analyzed Using ASP Methods, by Vendor

Docket/ LER No.	Description of event (date)	Conditional core-damage probability*
GENERAL ELECTRIC (BWR)		
271/89–013	10,000 gal (37.9 kL) of reactor vessel inventory was transferred to the torus at Vermont Yankee. (3/9/89)	1×10^{-6}
387/90–005	Extended loss of RHR at Susquehanna 1. (2/3/90)	3×10^{-5}
397/88–011	Loss of reactor vessel inventory at WNP–2. (5/1/88)	5×10^{-5}
458/89–020	15,000 gal (51.8 kL) of service water flooded the auxiliary building when a freeze seal failed at River Bend. (4/19/89)	1×10^{-6}
BABCOCK AND WILCOX (PWR)		
287/88–005	Loss of ac power and loss of RHR during midloop operation with vessel head on at Oconee 3. (9/11/88)	2×10^{-6}
302/86–003	RHR pump shaft broke during midloop operation at Crystal River 3. (2/2/86)	1×10^{-6}
COMBUSTION ENGINEERING (PWR)		
285/90–006	Loss of offsite power (LOOP) with the emergency diesel generators (EDGs) not immediately available at Fort Calhoun. (2/26/90)	4×10^{-4}
382/86–015	Loss of RHR during midloop operation at Waterford 3. (7/14/86)	2×10^{-4}
WESTINGHOUSE (PWR)		
323/87–005	Loss of RHR at Diablo Canyon 2 while in midloop operation. (4/10/87)	5×10^{-5}
456/89–016	RCS inventory loss at Braidwood 1. (12/1/89)	1×10^{-6}

*See Section 2.2.3 for the limitations to this approach.

Probability values estimated using these approaches are very uncertain. Unfortunately, these same probabilities significantly influence the conditional core-damage probabilities estimated for the two more significant events and, therefore, those conditional probabilities are also uncertain.

The impact of long-term recovery assumptions is illustrated below. Changes in conditional probabilities resulting from a factor-of-three change in the non-recovery estimates are listed for the Susquehanna and Waterford events. As can be seen, within the range shown, the conditional probability for both events was very strongly related to assumptions concerning long-term recovery.

Operator response is probably the most important issue determining the significance of an event in shutdown, and until it is better understood, the relative importance of shutdown events compared to events at power cannot be reliably estimated.

2.2.3 Results and Findings

The conditional core-damage probabilities estimated for each event are listed in Table 2.7 and illustrated in Figure 2.1. The calculated probabilities are strongly influenced by estimates of the likelihood of failing to recover initially faulted systems over time periods of 6 to 24 hours. Very little information exists concerning such actions; hence, the conditional probability estimated for an event involved substantial uncertainty. Additionally, some conditional probabilities were strongly influenced by assumptions concerning (1) the plant staff's ability to implement non-proceduralized short-term actions, (2) the actual plant status at the time of the event, and (3) the potential for the event to have occurred under different plant conditions. The distribution of events as a function of conditional probability is shown in Table 2.9. The result for the 1990 loss of ac power and SDC at Vogtle 1 is also included for completeness. The analysis performed for the Vogtle 1 event is documented in NUREG/CR–4674, Volume 14. Events with conditional probabilities below 1×10^{-4} are considered minor with respect to risk of core damage. Conditional probabilities above this value are indicative of a more serious event.

Excluding the Vogtle loss-of-all-ac-power event, the two events with conditional probabilities above 10^{-4} are:

(1) *Loss of Offsite Power With an EDG Out of Service at Fort Calhoun on February 26, 1990.* During a refueling outage, a spurious relay actuation resulted in isolation of offsite power supplies to Fort Calhoun. One diesel generator (DG) was out of service for maintenance, the other started but was prevented from connecting to its engineered safety features (ESF) bus by a shutdown cooling pump interlock. Operators identified and corrected the problem, and the DG was aligned to restore power to the plant. The conditional probability of core damage estimated for this event is 3.6×10^{-4}. The dominant sequence involves failure to recover ac power.

The calculated probability is strongly influenced by estimates of failing to recover ac power in the long term. These estimates involve substantial uncer-

tainty, and hence the overall core damage probability estimated for the event also involves substantial uncertainty.

(2) *Loss of Residual Heat Removal (RHR) During Midloop Operation at Waterford 3 on July 14, 1986.* In this event, a non-proceduralized drain path was not isolated once the reactor coolant system (RCS) level was reduced to midloop. Draining continued and resulted in cavitation of the operating RHR pump. Restoration of shutdown cooling (SDC) took 3 hours, during which time boiling occurred in the core region. Both RHR pump suction lines from the RCS were steam bound (most likely a result of the suction loop seal design feature). RCS inventory was restored using one of the low-pressure safety injection (LPSI) pumps (these are the same as the RHR pumps on this plant) taking suction from the refueling water storage pool.

Shutdown cooling was eventually restored by using the pump warmup lines in conjunction with repeated pump jogging—a non-proceduralized action. The method specified in the procedure to restore RHR pump suction (use a vacuum priming system to evacuate the loop seal) would not have been effective since hot-leg temperature exceeded 212 °F (100 °C).

The dominant core-damage sequence for this event (which includes the observed failures plus additional postulated failures, beyond the operational event, required for core damage) included an assumed failure to recover RHR, in combination with an assumed unavailability of the steam generators as an alternative means of removing decay heat.

Table 2.9 Events Listed by Conditional Core-Melt Probability

Conditional probability range	Description of event
10^{-3}	Loss of all ac power at Vogtle (NUREG–1410)
10^{-4} to 10^{-3}	Loss of offsite power with EDG out of service at Fort Calhoun (LER 285/90–006)
	Loss of RCS inventory and SDC during midloop operation at Waterford 3 (LER 382/86–015)
10^{-5} to 10^{-4}	Loss of RCS inventory and SDC during midloop operation at Diablo Canyon 2 (LER 323/87–005)
	RHR isolation of Susquehanna 1 (LER 387/90–005)
	Loss of RPV inventory at WNP–2 (LER 397/88–011)
10^{-6} to 10^{-5}	2 events considered minor with respect to risk of core damage
10^{-6}	3 events considered minor with respect to risk of core damage

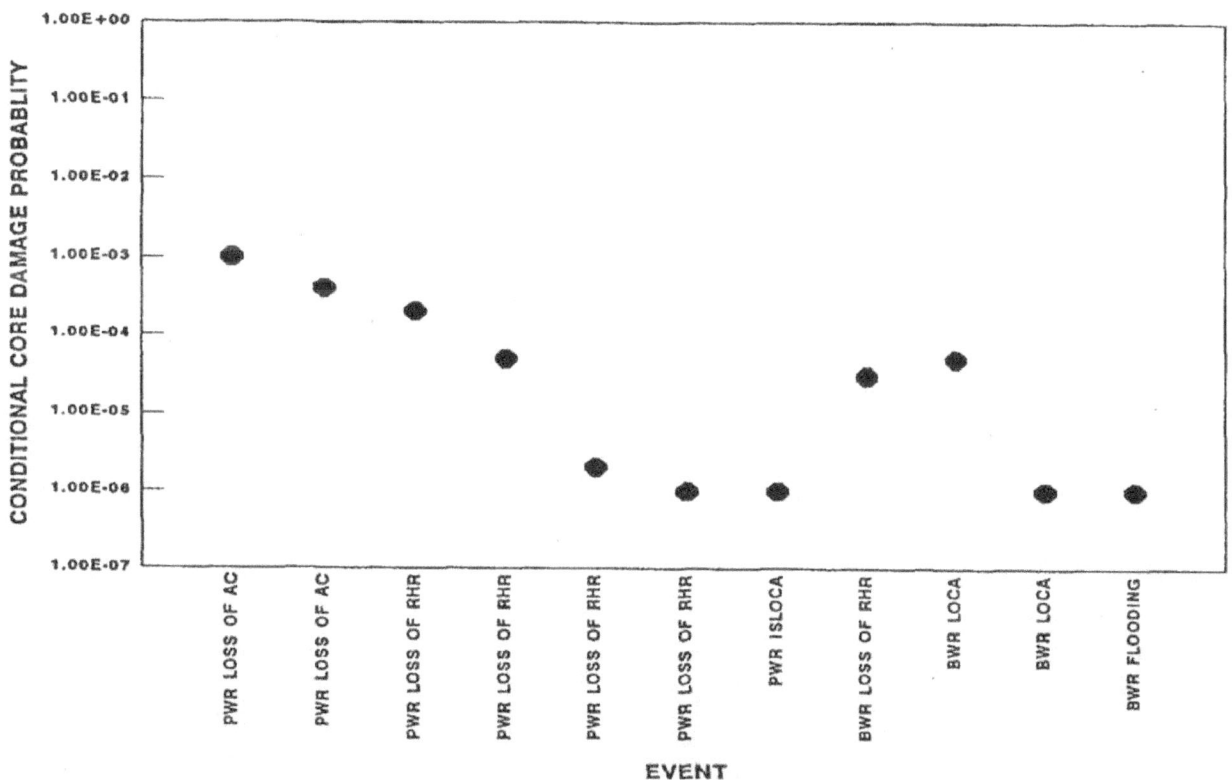

Figure 2.1 Accident sequence precursor results

One significant common factor that resulted in the higher conditional probability estimates for these events was the inability to passively drain water from the refueling water storage tank to the reactor vessel because there was no elevation head. Key factors that impacted risk estimates for many of the events treated in this study are discussed below, along with other analysis findings.

2.2.3.1 Design and Operational Issues Important to Risk During Shutdowns

Plant Procedures. Procedures in use at the time of the event had a significant effect on the analysis of the event, since what operators knew about alternative recovery strategies was assumed to derive primarily from the procedures. Ad hoc actions were postulated in some cases, but were considered much less reliable than proceduralized actions. Detailed guidance was limited in early procedures, and what did exist offered little information on how to recognize an event or implement a correct recovery course. Some procedures did direct operators to substitute systems if RHR could not be recovered, but information needed for determining when such systems would be effective (such as the minimum time after shutdown before the system could adequately remove decay heat) was not given.

Contemporary procedures offer much greater guidance and flexibility, both in the number of substitute systems that can provide RHR and in information to help characterize an event. For example, Crystal River 3 now has a procedure specifically directing the operators to use five different systems for makeup water, whereas in 1986 (when the event analyzed in this study occurred), the procedures listed only two such systems. The current loss-of-RHR procedure for Braidwood lists seven other methods to reestablish core cooling, gives tabular guidance regarding which methods are effective for different operating states, and provides graphs as a function of time since shutdown for RCS heatup required vent paths and required makeup flow for RHR.

If events similar to those analyzed in this report occurred now, many would be considered less significant from the standpoint of risk of core damage because of the additional guidance and flexibility now included in the procedures.

Operator Recovery Actions. Differences between operator actions associated with recognizing that an event was in progress, detecting the cause of a problem, and implementing recovery actions are apparent in the descriptions of many of the 10 events. Several events were taking place for some time before someone either recognized there was a problem or was able to identify its exact nature. For example, during the Vermont Yankee event, operators took 15 minutes to recognize that the water level in the reactor vessel was decreasing and then they spent the next 30 minutes determining the source of the leak. Once it was found, the source of the leak was quickly isolated.

For the event at Braidwood, operators quickly concluded that an RHR suction relief valve had lifted. However, 2-1/2 hours were required to locate the valve that had lifted (it was on a non-operating train).

For both the Vermont Yankee and Braidwood events, SDC was not lost and a lot of time was available to detect and correct the problem before core cooling would have been affected. This was important, because it gave the operators time to deliberately and systematically address each event. Availability of a long time period before the onset of boiling or core uncovery was reflected in lower probabilities for failure to recover a faulted system or implement actions away from the control room.

On the other hand, in the Waterford event (which happened when SDC was lost during midloop operation), boiling initiated approximately 45 minutes after SDC was lost. This is a short period of time to reliably implement recovery actions out of the control room. For the loss of SDC at Waterford, information concerning RHR pump restart (use of the vacuum priming system to evacuate the suction lines) was not correct for the RCS condition that existed during the event. SDC was eventually restored by repeated pump jogging and the use of pump warmup lines to return some flow to the pump suctions.

Design Features That Complicate Recovery of RHR. The loss of SDC at Waterford illustrates a design feature that significantly affected recovery of SDC. At Waterford, loop seals exist in both the RHR suction and discharge lines. The loop seals are more elevated than the RCS loops and the top of the refueling water storage pool (RWSP). During the 1986 event, SDC suction flow could not be quickly restored, because of steam in the shutdown cooling system. For that event, the procedure for responding to loss of SDC did not adequately address all RCS conditions that could be expected following a loss of SDC, nor did it provide information on plant features that could complicate recovery. (Although not important in the recovery of the 1986 event, the loop seals would also prevent the use of gravity feed from the RWSP for RCS makeup.)

Diverse Shutdown Cooling Strategies. The availability of diverse SDC recovery strategies can play a significant role in reducing the significance of events. Use of a diverse system to recover SDC would not require the recovery or repair of an initially faulted system, and presumably could be implemented more quickly in many cases.

Many of the new procedures identify diverse methods for RHR. For example, the Braidwood procedure regarding loss of RHR identifies the following alternate core cooling methods:

- bleed and feed using excess letdown through loop drains and normal charging

- steaming intact/non-isolated steam generators

- bleed and feed using pressurizer power-operated relief valves

- refuel cavity to fuel pool cooling

- safety injection pump hot-leg injection

- accumulator injection

- inventory addition via the refueling water storage tank.

Not all of these methods are applicable at all times; however, they offer a significantly greater flexibility than a procedure in which just one alternative method is specified in addition to recovery of the faulted RHR system.

2.2.3.2 Factors That Strongly Influence the Significance of an Event.

Analysis of the 10 events confirms the influence of a number of factors on significance. These factors are described below.

High Decay Heat Load. A high decay heat load significantly reduces the time available for SDC recovery before boiling or core uncovery. This, in turn, increases the probability of failing to recover SDC or implementing alternate cooling strategies, and may also increase the stress level associated with the event. The number of alternate systems that can effectively remove decay heat is also fewer than at low decay heat loads; that may further complicate recovery.

RCS Inventory. Having the refueling cavity filled with water to a level [23] feet with upper internal equipment removed increases the time available for SDC recovery significantly with a similar impact on the reliability of operator actions. In contrast, midloop operation in a PWR is performed with minimal RCS inventory, and by its very nature decreases the reliability of the RHR system.

Status of Reactor Vessel Head. Events that occur when the head is removed are typically less significant than those that occur with the head on, since RPV makeup combined with core region boiling will provide RHR.

Availability of Diverse Systems for SDC. The availability of diverse systems that can operate independently of components in the RHR system reduces the risk associated with a loss of SDC, since availability of these systems does not depend on recovery of the RHR system.

Adequate Procedures. Procedures that give detailed information concerning response to a loss of RPV inventory or SDC, and alternate strategies for recovery, are important.

3 SITE VISITS TO OBSERVE SHUTDOWN OPERATIONS

Small teams of NRC personnel, each comprising from 2 to 4 technical people, observed low-power/shutdown operations at 11 nuclear power plant sites during 1991. The teams' main objectives were to observe plant operations during shutdown and learn about the policies, practices, and procedures used to plan outage activities and conduct them safely. The teams' observations, supplemented by data obtained from recent NRC inspections at six other sites, are presented in this chapter. At the 17 sites, 29 units were operating—4 Babcock and Wilcox, 5 Combustion Engineering, 6 General Electric, and 14 Westinghouse.

On the average, a team spent about a week at a site during an outage. During that period, the team interviewed all levels of utility personnel and observed activities taking place in the areas of operations, management, and engineering, including daily meetings of the plant staff to assess progress and problems concerning the outage work in progress.

3.1 Outage Programs

Programs for conducting outages varied widely among the sites visited.

Susquehanna's program for conducting outages was among the best. It included (1) prudent, practical, and well-documented safety principles and practices; (2) an organization dedicated to updating and improving the program as well as monitoring its use; (3) strong technical input to the program from the onsite nuclear safety review group; (4) a controlled program manual concurred in by line management and familiar to appropriate personnel; and (5) training on the program and the program manual.

Another site that was visited had no comparable program and was poorly prepared and poorly organized, which was reflected by failure to complete planned work in past outages, long outages, and by the team's other observations of work in progress. At several plants, licensees had neither documentation nor plans to provide any. Two plants made exceptional efforts to keep outages short. At one of these two plants, the team noted examples of less prudent operation than at other plants it visited. The other plant had a greater number of recent shutdown-related events than any plant visited.

3.1.1 Safety Principles

Well-founded safety principles play a significant role in an outage program. Sites visited varied widely in this area. A high priority was seldom placed on such principles, and sometimes safety was based upon individual philosophies.

Often, principles were "understood" in contrast to being clearly defined in a documented management directive.

Some licensees emphasized safety in outage planning and during outage meetings. They posted critical safety boundaries at key locations and identified and tracked critical safety equipment with as much emphasis as given to critical path. Some pressurized-water-reactor (PWR) licensees were particularly sensitive to midloop and reduced inventory operation. One site presented the following good safety principles in its program:

- Minimize time at reduced inventory.

- Maximize pathways for adding water to the reactor coolant system (RCS).

- Maximize availability of important support systems.

- Minimize activities requiring midloop operation.

- Maximize time with no fuel in the reactor vessel (RV).

Some sites visited gave in-depth consideration to such safety areas as criticality, containment, instrument air, electric power, gravity feed, steam generator (SG) availability (in case of RCS boiling), use of firewater, and other areas. Others relied upon an ad hoc approach should problems arise.

3.1.2 Safety Practices

A wide variety of safety practices was noted. Some utilities adhered to a "train outage" concept, removing an entire train, including electrical equipment, pumps, controls, and valves, from service. The other train was "protected," no work was allowed on it. Stated benefits were avoidance of train swaps, minimization of mistakes, and simplification of the operator's job. A "block" approach was also used in which a boundary was established and work was allowed within that boundary as long as no water was moved. Other utilities practiced different approaches that may allow more flexibility, but placed greater dependence on their personnel to avoid conflicts. Other safety practices observed by the team included the following:

- Provide sufficient equipment that no single failure of an active component will result in loss of residual heat removal.

- Add one injection system or train to that required by technical specifications (TS).

- Provide multiple power supplies, batteries, charging pumps, and such.

- Always have one emergency core cooling system (ECCS) available.

- Comply with TS; these are sufficient to ensure safety.

3.1.3 Contingency Planning

Some licensees provided in-depth preparation for backup cooling, whereas others placed more reliance on ad hoc approaches. Backup cooling includes such techniques as gravity feed, allowing RCS boiling in PWRs with condensation in SGs, and use of firewater. Again, there were many variations in both capability and planning. Some PWR licensees planned SG availability; others did not. Some who planned for the use of firewater and staged spool pieces had procedures; others did not. Most PWRs had some gravity feed capability during some aspects of shutdown operation; others did not. Those that did may or may not have had good coverage in procedures. No site visited had planned ECCS accumulator usage. All of these capabilities are potentially important and could effectively terminate many events.

3.1.4 Outage Planning

Planning ranged from initiating work a few months before an outage was scheduled to having plans that covered the life of the plant, including anticipated license extensions. There was evidence that good planning, including experience, averted many outage difficulties. Conversely, poor planning appeared to be a cause of such outage difficulties as extended schedules and failure to complete work.

The following items provide additional perspective regarding planning adequacy and effectiveness:

- Well-planned and tightly controlled outage plans allowed for increase in the scope and number of unanticipated activities that seldom exceeded 10 to 20 percent. Conversely, growths of 40 percent and more than 100 percent correlated with outages that lasted longer than planned, that were poorly managed, and that sometimes resulted in a return to power with significant work unaccomplished.

- Some licensees could enter an unscheduled outage and have a complete outage plan within hours. Others had no bases and worked only on the item causing the shutdown. In one case, a licensee entered a refueling outage a month early but accomplished little work before the originally scheduled start date. Another licensee entered a refueling outage a month early, moved the completion date up, and completed the outage in the original time allotted (a month early when compared to the original plan).

- In smaller, less-complicated plants, highly experienced licensee staffs could conduct apparently well-coordinated refueling outages with only a few months of planning. Key contributing factors appeared to be having few inexperienced people, having the experience of many refueling outages, having a good plan that was prepared quickly, and anticipating material needs well in advance of preparing the plan. Some other licensees, both experienced and relatively inexperienced, had what were judged as relatively poor plans, and their outages appeared to be in some disarray. Finally, some licensees with few refueling outages were able to conduct outages on schedule when they had good plans.

3.1.5 Outage Duration

Safety criteria and implementation effectiveness appeared to be more important to safety than outage duration. Refueling outage durations beyond roughly two months did not appear to increase safety. Conversely, a less-prudent safety approach may be instrumental in shortening outages. However, outage duration was also a function of plant type, the work to be done, planning, and implementation. A short outage was not necessarily an outage in which safety has been reduced to shorten the outage, although shortness was an indicator that one should look closely to see how the short schedule was achieved.

The teams observed that several licensees felt pressured to reduce outage time further than the team judged to be prudent. Reasons given included being rated by others on the basis of a short outage time and being driven toward a fuel critical path to shorten outage time.

Numerous approaches to planning affected outage time, including the following:

(1) Do not reduce refueling outage time below a somewhat judgmental minimum because safety might be jeopardized (several licensees). Typically, these licensees applied safety criteria throughout the outage and these criteria sometimes determined critical path.

(2) Define one critical path, such as the refueling floor, and normally force everything else to fit.

(3) Allow critical paths to float depending upon the work schedule. Safety considerations may influence critical path. (Often, items 1 and 3 were followed simultaneously.)

(4) Describe the work and suggest schedules to "corporate headquarters." Receive or negotiate an allowable outage time.

3.1.6 Outage Experience

All licensees incorporated outage experience into planning and found feedback useful. Most provided for feedback during an outage. Some conducted team meetings immediately after completing significant tasks; others met following the outage. Most compiled outage reports and used these in planning the next outage. Typical results included the following:

- Place personnel with operations backgrounds into key positions and areas for planning and conducting outages.

- Locate the outage control location ("war room") close to the control room (CR) to facilitate communication.

- Assign a senior reactor operator who is adjacent to the CR, but not actually in it, to handle the work orders.

3.2 Conduct of Outages

Typically, outages were conducted with a licensed person who controlled tagouts and approved each work package before initiating day-to-day work. The daily (and other) outage meetings also provided an opportunity for identifying issues. Beyond this, various approaches were used, ranging from individuals who had their own criteria to various depths of written and unwritten guidance or criteria.

Some licensees were protective of critical equipment and made sure everyone was sensitive to such issues. For example, one licensee protected the operable train of safety equipment by roping off the areas and by identifying the operable train on every daily plan. Similar approaches to the protected train (including identifying it in the daily meetings) were found at several plants. Other techniques included providing critical plant parameters in the control room.

Licensees often changed their organizations for an outage, although some operated by incorporating shutdown features into the organization used for power operation and made few actual organization changes. There was a general trend to emphasize operations experience for outage positions at all levels. Licensees who had emphasized such experience considered it to be very beneficial in conducting a satisfactory outage.

Significant variations existed among sites visited in the ratio between utility manpower and total manpower, and in the percentage of personnel involved in the previous outage. Utilities that had a high percentage of people experienced in previous outages at that facility considered such experience to be a significant benefit. Among advantages cited were familiarity with the plant, less training, higher quality, shorter outages, and better motivated people.

Some licensees used task forces and "high impact teams" for critical-path and near-critical-path tasks. These groups were composed of experienced personnel who had performed the same function in past outages.

Contractors were used to various depths by different licensees. Their capabilities, licensee supervision, and influence on outages varied widely. Licensees who worked closely with their contractors and supervised them closely appeared to get better results than those who neither carefully trained nor supervised their contractors. Previous contractor experience at the site was often stated to be an advantage and licensees often tried to use the same contractor from outage to outage.

Interestingly, a large plant staff did not translate into an effective outage, nor did a smaller staff at a "small" plant translate into an ineffective outage. Staff size also did not necessarily correlate with safe operating practices, although the teams did encounter areas that were weak because they lacked manpower. Those plants judged to have the most effective safety programs were adequately staffed in areas directly related to safety, were well organized overall, and appeared to conduct effective outages.

All utilities conducted periodic reviews during outages. Typically, these involved overviewed specialized meetings that were held once or twice a day and involved all levels of plant personnel and all disciplines. All utilities provided computer-generated outage schedules in several formats and updated some of these every day (or more often). Schedules typically covered a day, 3 days, 7 days, and the complete outage, and provided a breakdown ranging from an overview through complete scheduling of all activities. Critical-path scheduling was seen often. Some utilities noted safety information prominently on their schedules; others did not.

Most daily meetings appeared well focused and to the point. Achievement appeared to vary widely. Most expectations were routinely met at some plants, but at others the outage appeared to be in disarray.

A commonly applied test for a satisfactory outage was meeting or bettering the outage schedule. Corollary tests were: (1) meeting ALARA (as low as reasonably achievable) goals, (2) avoiding personnel injuries, (3) completing planned work, (4) not having to repeat work during power operation (because it was done well during the outage), and (5) not having reportable events.

3.2.1 Operator Training

Licensees often conducted extensive training immediately before a scheduled outage, a practice judged necessary by most licensees because of the specialized nature of, and the lack of everyday exposure to, low-power and shutdown (LPS) operation. This was not always done, however, and minimal training was evident at some sites.

Some operators and instructors said they thought LPS operation was important, but that the NRC had implied otherwise by not emphasizing it more in exams and evaluations. Others felt that strong NRC interest in training was reflected in Generic Letter (GL) 88–17 inspections and independent resident inspector followup. Although GL 88–17 coverage was limited, licensees have applied the information to a wider range of PWR plant conditions.

LPS operations training was often specialized. Some licensees gave concentrated study in unique aspects of the outage to the operating shift expected to handle those aspects of the outage. Training often involved specific equipment, such as valves, reactor coolant pump seals, and steam generator (SG) manways. Capabilities such as a control rod handling machine mockup for a boiling-water reactor (BWR), SG plena mockups, valves, pumps, and an emergency diesel generator (EDG) model for maintenance training were observed.

As in many other areas, the quality and scope of training were varied, and ranged from

> Outage training is completed before the outage. Training for power operation with simulator upgrades is conducted before leaving the outage. Special tests are addressed as are evolutions, primary manway and nozzle dam work, level indication problems, procedures, and consequences of what can happen. Procedure changes, including background, are covered before crews take the watch.

to

> Many plant operators have not had overall systems training for several years and have had no formal outage-specific training since the initial response to GL 88–17.

3.2.2 Stress on Personnel

Although the teams considered stress in general, it was investigated in depth at only one plant. This licensee emphasized short outages, and operators perceived their achievement as related to outage time. Four operators (of seven interviewed in depth) said the outages were too short. Much of the direct outage coordination was conducted from the CR, which was smaller than many multiple-unit CRs. In many instances, such activities appeared to affect plant operation. Further, all operators said the

work load was high or very high. Operators also said they met the schedule with difficulty, that they sometimes took on more work than they could handle, that they had to cut corners to stay on schedule and then had to make repairs later, that they wrote procedures at the last minute in the CR, operated without some procedures, and had poor procedures for shutdown; all of the seven operators interviewed said they were poorly trained or that they had significant reservations regarding training. There were many other similar comments. All seven operators said stress was self-generated, and six also identified stress caused by pressure from non-operations personnel. Four operators said stress was severe enough to be a problem. These operators were working four 12-hour shifts followed by a break. No operator stated working hours were too long or that working hours contributed to a problem. This plant was judged to have significant operator stress problems that were reflected in numerous mistakes.

3.2.3 Technical Specifications

No TS were applicable during much of a refueling outage at one site as long as temperature measured at the residual heat removal (RHR) pump remained below 140 °F (60 °C) or 200 °F (93 °C), depending upon the interpretation. (Note that this temperature is unlikely to increase if the RHR pump is not running.) Another site had no TS on EDGs, batteries, and service water during shutdown operation. No plant visited had complete TS coverage.

Most of the industry stated that TS did not fully address LPS operations. The single exception reported that it planned outages on the basis of TS, and this was sufficient to ensure safety. Many personnel commented that existing TS were more appropriate to power operations than to LPS conditions.

Similarly, licensees were concerned with TS that caused extra work, resulted in extra dose, and sent an undesirable message to plant personnel. One example cited was the requirement for an operational pressurizer code safety valve although large openings existed in the RCS. The licensee estimated several hours of work and 500 mrem (5 mSv) of dose were involved to unnecessarily install and then remove the valve.

3.3 Plant and Hardware Configurations

The teams observed that configurations of plant systems and components used by licensees during outages varied widely among plants visited. During the visits, the teams examined configurations of equipment throughout the plants, including regions outside the protected area. The teams' observations in selected areas are presented below.

3.3.1 Fuel Offload

The fuel at some units was regularly offloaded; at some, fuel may or may not be offloaded. The fuel at other units would be offloaded only if there was no reasonable alternative.

An often-cited safety advantage for offloading was flexibility available because no fuel was in the RV, and the associated decrease of mistakes leading to a fuel cooling concern. Other considerations included loss of fuel pool cooling, flexibility in providing fuel cooling if systems were lost, fuel storage volume heatup rate upon loss of cooling, criticality, reduced operator stress due to avoidance of such conditions as midloop operation, and the potential to damage fuel during handling. Fuel offload had a significant advantage in that an early midloop operation, and sometimes all midloops, can be avoided, although not all licensees who offloaded also avoided an early midloop operation.

Several licensees performed an incore fuel shuffle and reported they encountered no problems with moving fuel within the core. They said that a complete core offload would lengthen their outages. Conversely, several licensees (both PWRs and BWRs) routinely performed a complete core offload, which they said was safer and provided more flexibility. Several licensees reported the offload path was faster than, or at least as fast as, an incore shuffle. Others offloaded or not on the basis of the planned outage work. Some decisions were based upon such considerations as the configuration (offload appeared to be difficult in Mark III BWRs), fuel distortion history, gains achievable with no fuel in the RV, and the reliability of the fuel handling machine.

3.3.2 Midloop Operation in PWRs*

Concerns about midloop operation appear to have influenced outage planning at many sites, but not at others. The team observed licensees who

- do not enter midloop operation under any circumstances.

- do not permit early midloop operation and defueling before installing nozzle dams.

- apply special midloop criteria to refueling outages, but deviate for an unscheduled outage

- routinely enter midloop within a few days to a week of power operation.

Some licensees required an additional operator in the control room for midloop operation. Another, whose hardware was particularly sensitive, required three additional operators who had specific responsibilities in the conduct of reduced-inventory operations; that is, opera-

tion when the RV water level is lower than 3 feet below the RV flange.

3.3.3 Venting in PWRs

RCS vents were sometimes of insufficient size, being smaller than planned and smaller than required by licensee procedures. Licensee personnel who recognized the implications were often unaware of these conditions.

Some licensees provided an RCS vent by removing one or more safety valves from the pressurizer. Others removed a pressurizer manway. If boiling develops, significant backpressure can occur from friction in the surge pipe, water traps, and the elevation head of the water held up in the pressurizer. Licensee personnel did not always recognize these phenomena.

Licensee personnel usually used covers or screens to keep foreign material from falling into pressurizer openings. These were often makeshift installations that could cause additional backpressure. Most licensee personnel interviewed by the team were unaware of the covers or screens.

The staff has identified some licensees who rely on lifting of the reactor pressure vessel (RPV) head on detensioned bolts for vent capacity during operation with a reduced inventory. In this approach, water is supplied from the refueling water storage tank (RWST) at a flow rate sufficient to support subcooled decay heat removal and to lift the reactor vessel head less than an inch, allowing water to spill over the vessel flange. The flow of water into the vessel is throttled with a flow control valve to prevent the head from lifting off the upper internals of the core. This is important because as long as the head rests on the upper internals, the internals' alignment pins will prevent it from cocking. This method works only when the decay heat load is low enough so that subcooled decay heat removal can be accomplished without lifting the head off the upper internals. Subcooled decay heat removal is necessary because venting steam past the vessel head can result in nonuniform head lift (cocking) and damage to the head due to cyclical impact loads. The acceptability of using the "head-lift" method for venting during operation with a reduced inventory depends on a number of plant-specific factors which should be thoroughly evaluated with appropriate analysis in the areas of thermal-hydraulics and engineering mechanics.

3.3.4 Nozzle Dams* in PWRs

Some PWR plants use nozzle dams and some do not. The recent trend in Babcock and Wilcox nuclear steam supply systems has been to use them, whereas a few years ago this was seldom done. One licensee attributed outage savings

---

*A midloop condition exists whenever RCS water level is below the top of the flow area of the hot legs at the junction with the reactor vessel.

*Nozzle dams are temporary seals installed in RCS primary piping that isolate components such as steam generators from reactor vessel and reactor cavity water so that work can be done on the components.

of close to a week to the use of nozzle dams, whereas another had them but did not use them and typically spent 3 to 14 days at midloop. Others indicated they might be at midloop for close to a month without them.

One licensee indicated there was no analysis to cover midloop operation with both nozzle dams and the RV head installed and such operation would not be permitted until the analysis was completed. The team noted that this observation was similar to others regarding incompleteness of analyses of shutdown operation.

3.3.5 Electrical Equipment

An outage typically represents times when equipment unavailability is high, unusual electrical lineups exist, and the likelihood of an electrical perturbation is increased by maintenance activities. The teams identified several events that could lead to electrical component damage or loss at some facilities, and concluded that almost all of those identified events could be easily eliminated. The team also found that protection and control of offsite electrical power systems varied.

Approaches to provide ac power included the following:

- Allow cooling via a system powered by a non-safety-related bus with no procedures for providing safety-related power to that bus.

- Provide one EDG and one source of offsite power.

- Provide one less source of power during shutdown to allow maintenance on one source at a time.

- Always have three sources of power, one of which is an EDG. (The site that advocated this did not have an EDG for about 2 weeks with fuel off-loaded, but it had a temporary diesel available.)

- Have both EDGs operable when in midloop operation. (One licensee stated it did not consider it prudent to stay at midloop conditions with only one EDG and would leave midloop operation if the second EDG could not be made operable quickly.)

- Allow both EDGs to be out of service when the fuel is offloaded.

- For midloop operation, normally have two EDGs and two offsite sources and allow no battery work, no reserve auxiliary transformer outage, no work that affects safeguards buses, or anything that affects the RCS. Otherwise, always require two off site and one on site.

- Make at least three separate ac power sources available to the vital buses any time two RHR pumps are

required to be operable. In practice, one of the sources has to be an EDG.

Additional variations include switchyard restrictions, restricting work on, or access to, vital areas such as near an operable EDG or operable electrical equipment, information requirements, administrative procedures, and whether variations are permitted and what level of management is necessary to approve such variations.

EDG maintenance and associated testing are usually performed during shutdown, although some licensees were performing this work at power. Also observed was removal of an EDG from service via entering Action statements immediately before shutdown.

Concerns also involved whether to have EDGs operating or operable. Potential decreases in EDG reliability due to grid disturbances and other perturbations, extensive testing, and running with a small electrical load were identified as potential problems with having EDGs operating.

Most plants had transformers and often breakers within the site's protected area. Switchyards were located nearby, but usually in whole or in part outside the protected area. These switchyards may contain a few transformers, but often contained only breakers and switches. They were usually fenced if outside the protected area, and usually had a locked gate. Often there was a control building within the switchyard, with attendant vehicle traffic. This building was seldom located adjacent to a switchyard entrance gate.

The teams did not observe any evidence of vehicle impacts within switchyards. However, they did find such evidence on both transformers and supports located within unfenced areas within site-protected areas; they also found a number of damaged fences. In one case, the source of safety-related offsite power entered the turbine building roughly 1 foot from where heavy trucks and trailers were sometimes parked, and was protected only by an ordinary chainlink fence. Fire hydrants at all sites were protected by a profusion of concrete-filled pipes, but at many sites important transformers within a few feet of the hydrants were unprotected. Switchyards were typically full of towers and bus supports. Some of the weakest supports were located in the corners and typically supported ring buses—loss of which could cause a loss of offsite power. These corner towers were often the towers most exposed to traffic within the switchyard, yet they were unprotected.

Some sites maintained CR control over switchyards outside the site's protected area. Other switchyards could be entered by anyone who had a key to the padlock; often, a utility staff member not assigned to the nuclear facility had a key, and sometimes someone who was not even an employee of the same utility had a key. Sometimes control was provided if the plant was in a sensitive condition, such as a PWR in midloop operation, but at other sites switch-yard work could proceed with little or no consideration of the nuclear plant status. At one plant, the team found the switchyard gate open and no one monitoring traffic at the gate. This switchyard was in an uncontrolled area.

3.3.6 Onsite Sources of AC Power

Onsite sources of electric power that were observed included diesel generators, hydro units, and portable power supplies. The most common source of safety-related power was EDGs.

Many variations in EDGs and configurations were seen. Size ranged from a fraction of a megawatt to 8 MW. One two-unit plant had two EDGs and routinely performed maintenance on one EDG while one unit was at 100-percent power and the other was in a refueling outage. That site planned to add two more diesels. In contrast, the Susquehanna two-unit plant had five EDGs. The fifth could be used as a complete replacement for any of the other four with no difference in CR indication and plant operation. Susquehanna also provided a portable diesel for battery charging and other uses should all ac power be lost for an extended time.

Roughly a third of the plants visited had the capability to resupply the EDG starting air tanks without ac power. The dominant method was a single-cylinder, diesel-powered compressor; but instrument air, a cross-connect with another EDG's air supply, and changing the drive belt from the electric motor to a one-cylinder engine were also observed.

3.3.7 Containment Status

Some PWR licensees closed the containments for conditions other than refueling; others did not, unless they entered a condition as described in GL 88-17. Some did not remove their equipment hatches during routine refueling outages; others did. Some provided containment closure capability that would withstand roughly the containment capability; others could lose containment integrity at roughly 1 psi. Some had proven containment integrity; others did not, and may not have attained an integral containment that meets GL 88-17 recommendations.

BWR secondary containments were judged unlikely to prevent an early release following initiation of boiling with an open RCS or during potential severe-core-damage scenarios. Among the BWRs, only the Mark III primary containment appeared potentially capable of preventing an early release without hardware modifications during such events. See Section 6.9 for a more complete assessment of containment capability. In general, no plans were found in BWRs for containment closure or for dealing with conditions under which the containment may be challenged.

3.3.8 Containment Equipment Hatches

A majority of the equipment hatches seen at PWR sites can be replaced without electrical power. See Section 6.9.3 for a full discussion of equipment hatch design and operation. It appeared that many licensees failed to check for adequate closure as addressed in GL 88-17.

The team learned that Arkansas Nuclear One had a requirement that an equipment hatch be capable of closure within approximately 15 minutes of a loss of RHR. Responsibilities were established for such actions as notification of loss of RHR, containment evacuation, closure operations, and verifications. Tools were kept in a closed box at the hatch and were clearly labeled "for emergency use only." Unannounced closure exercises had been conducted. Few other sites visited were as well prepared.

A common weakness was failure to check for adequate closure. GL 88-17 specified "no gaps," not the "four bolts" commonly observed. The four-bolt specification appeared to be insufficient at some plants with inside hatches (hatches that would be forced closed by containment pressurization).

Oconee provided a small standby generator in case ac power was lost. This could be immediately used to power the winches that normally raise and lower the hatch. This appeared to be an excellent approach to one of the problems of loss of ac power.

3.3.9 Containment Control

Some licensees carefully controlled containment penetrations during LPS operation. Others were concerned only with TS requirements regarding fuel movement and reduced inventory/midloop commitments in their response to GL 88-17. Provisions were found to bring services such as hoses and electrical wires into the containment via unused containment penetrations at several sites. Such provisions made it easier to close the equipment and personnel hatches. Some licensees simply removed a blind flange and passed wires or hoses through the opening. Others provided a manifold arrangement that may effectively eliminate most of the open penetrations. Occasionally, a permanent connection or an adaptation of a penetration such as was used for containment pressurization was found for introducing temporary utilities. U-pipes filled with water were observed in use as a containment penetration seal. These were judged to be of little value in protecting against an accident involving significant steam production or a core melt.

A number of licensees planned to initiate containment closure immediately upon loss of RHR. Others were less stringent, including such possibilities as initiating closure if temperature exceeds 200 °F (93 °C). That approach is likely to allow boiling before containment closure, and boiling may make it impossible to continue closure operations. In one case, the licensee assumed personnel could work inside the containment in a 160 °F (71 °C) environment while closing the equipment hatch. More detail on this topic is given in Section 6.9.4.

Knowledge of what must be closed and providing the resources to actually close the openings and/or penetrations under realistic conditions were often overlooked. Tracking openings, providing procedures, and conducting walkthroughs that accounted for reasonably anticipated conditions were seldom found.

3.3.10 Debris in Containment

Blocking a PWR containment sump with debris from outage work may prevent effective recirculation of reactor coolant following an accident during shutdown. For example, PWR emergency core cooling (ECC) sump screens were removed during refueling outages at some sites, and at others the screens were covered with heavy plastic sheeting. In one plant, one screen was removed and the other was 10-percent uncovered to allow a recirculation capability. In another, one sump was open and the other was closed. Similar conditions were seen in plants with ECC connections in the bottom of the containments without a sump. In one, both filters were removed to expose the pipe opening; in another, the filters were in place. Actual and potential debris existed at all of these sites, but was seldom considered with respect to recirculation capability during shutdown.

3.3.11 Temperature Instrumentation

Core temperature during shutdown in PWRs was obtained by measuring water temperature just above the core by means of thermocouples. Other temperature indications required an operating RHR system for accurate indication of meaningful RCS and core temperature over a wide span of RCS conditions. Although this was addressed in GL 88–17, many operators were still unaware of the potential error associated with lack of flow. In numerous PWR heatup events, no temperature indication was available, although the frequency is decreasing as licensees implement the recommendations of GL 88–17. However, the team often observed poor application of the temperature coverage recommendation, principally involving not providing temperature indications for extended periods of time, restricting the indication to reduced inventory conditions, and failure to provide suitable alarms. Licensees who emphasized temperature indication generally provided measurements while the head was on the RV, except for the 30 minutes to 2 hours just prior to removing the head.

BWR coolant temperature was obtained by measuring the RV wall temperature and assuming natural circulation in the RV. The natural circulation assumption is not valid if water level is lower than the circulation paths in the steam separator. This was often unrecognized, and BWRs have encountered significant heatup with no indication of increasing temperature provided to the operators.

3.3.12 Water Level Instrumentation

BWRs were equipped with multiple water level indications that were on scale during both power and shutdown operation. PWRs were often operated with all of the "permanent" level indications off scale or inoperative during shutdown. PWR licensees have added level instrumentation to cover shutdown operation in response to GL 88-17. As observed, instrumentation in the BWRs was generally superior to instrumentation in the PWRs. The team often found many damaged and/or incorrectly installed instrument tubes inside PWR containments. Only one short tube section with an incorrect slope was found in a BWR. Many personnel described problems with maintaining accurate level indication in PWRs. No one described this problem in BWRs.

BWR level systems typically used a condensing pot to ensure that connecting pipes remain full, yet no condensate is generated during shutdown. No one indicated this has led to level indication error, nor did anyone identify this as a potential problem.

PWR level indications have significantly improved in the last 3 years. All PWRs now indicate level on the control board. In-containment installations often (but not always) showed evidence of professional installation that was missing several years ago. Much less reliance was being placed on temporary tubing runs. Several licensees were still working to meet GL 88-17 recommendations.

Some PWRs were equipped with ultrasonic hot-leg and cold-leg level indications. A few have been in operation for years, and this indication has been used in foreign plants for some time. Most licensees appeared satisfied with indication accuracy and reliability, although problems were reported with equipment obtained from one vendor.

3.3.13 RCS Pressure Indication

RCS pressure indications were generally wide range and not appropriate for monitoring shutdown operation. A number of operations personnel stated that the computer provided monitoring and cathode-ray tube indications that were more sensitive.

3.3.14 RHR System Status Indication

GL 88-17 identified pump motor current, RHR pump noise, or RHR pump suction pressure for monitoring RHR operation in PWRs. Although many licensees have followed the recommendations in GL 88-17, some responses have been minimal. Among the weaknesses observed were failure to provide a sensitive means to monitor RHR pump operation, failure to consider sampling rate when monitoring parameters, failure to provide trending information, too wide a pressure range to permit observation of behavior, and RHR systems operating with temperature off-scale low.

3.3.15 Dedicated Shutdown Annunciators

Numerous control room annunciators were typically lit during shutdown conditions. Arkansas Nuclear One had installed an annunciator board that addressed major shutdown parameters and was making it operational—the only such panel observed. Several operators indicated that even grouping existing parameters into an easily recognized pattern would be better than what they have. Others said they were familiar with the lit annunciators and had no difficulty recognizing an unusual pattern.

4 PROBABILISTIC RISK ASSESSMENTS

Risks associated with shutdown and refueling conditions have not been extensively studied and are not as well understood as are those associated with power operation. Few studies address the full scope of understanding about shutdown risk in pressurized-water reactors (PWRs) and fewer address such risk in boiling-water reactors (BWRs). Several probabilistic risk assessments (PRAs), including the ongoing NRC-sponsored Grand Gulf and Surry shutdown studies (currently at a preliminary level 1 stage), are summarized here to identify significant issues and insights associated with activities at nuclear power plants during shutdown and refueling outages.

4.1 NSAC-84

NSAC-84 was an extension of the Zion Probabilistic Safety Study completed in 1981. Procedural event trees were developed to account for changes in plant conditions during shutdown. Human errors and equipment failures unrelated to procedures were also considered. The initiating events studied were loss of residual heat removal (RHR) cooling, loss-of-coolant accidents (LOCAs), and cold overpressurization (excess of charging, over-let-down, or an inadvertent safety injection). A shutdown database specific to Zion was developed from plant records and used in quantification.

Findings

The mean core-damage frequency (CDF) at shutdown was estimated to be 1.8×10^{-5} per reactor-year.

Examination of the top 10 core-damage sequences revealed the following:

(1) Failures during reduced-inventory operation (including equipment unavailabilities and operator errors) appear in eight sequences, totaling 61 percent of the total CDF; failure of the operator to respond during reduced-inventory operation appeared in five sequences, accounting for 44 percent of the total CDF.

(2) Since malfunctions of RHR components require some type of operator intervention, all shutdown core-damage scenarios (due to overdraining of the reactor coolant system, LOCAs, and RHR suction valve trips) are sensitive to the operator's failure to restore core cooling. The operator's failure to determine the proper actions to restore shutdown cooling appeared in six sequences, accounting for 56 percent of the total CDF.

(3) Loss of RHR cooling (primarily pump and suction valve trips) was the initiating event in eight sequences, totaling 56 percent of the CDF; a LOCA was the initiating event in the other two sequences, totaling 6 percent of the total CDF.

4.2 NUREG/CR-5015 (Loss of RHR in PWRs)

NUREG/CR-5015 was issued in response to Generic Issue 99 concerning the loss of RHR in PWRs during cold shutdown. This study used the NSAC-84 methodology (based on the Zion plant configuration) with several modifications which included the consideration of loss-of-offsite-power (LOOP) events using a separate event tree and the use of generic event frequencies from PWR experience over a 10-year period from 1976 to 1986.

Findings

The mean CDF at shutdown was estimated to be 5.2×10^{-5} per reactor-year, with the following breakdown by initiating event:

- loss of RHR   82%
- loss of offsite power   10%
- loss-of-coolant accident   8%

Examination of the findings reveals that operator failure to diagnose that a loss of cooling has occurred and to successfully restore it while at reduced inventory in the reactor coolant system (RCS) accounted for 64 percent of the total CDF. The two dominant core-damage sequences involved a loss of RHR pump suction as a result of over-draining of the RCS.

The findings of NUREG/CR-5015 appeared to correspond to those of NSAC-84. Operator errors dominated the risk, particularly during midloop operation. LOOP events contributed to 10 percent of the total CDF, a relatively small contribution.

4.3 Seabrook PRA for Shutdown Operation

The Seabrook PRA information was collected from a number of presentations the licensee made to the NRC. This study supplemented the level 3 Seabrook PRA by examining the likelihood of core damage for the plant in standard Modes 4 (hot shutdown), 5 (cold shutdown), and 6 (refueling). Radiological source terms and public health consequences were also considered. The approach used to model accident sequences was similar to that used in NSAC-84 with several enhancements which included the following: fire and flood initiating events unique to plant shutdown were quantified and considered, an uncertainty analysis of the results was performed, the PWR experience database from NSAC–52 was updated and examined with insights being incorporated into plant shutdown models, and thermal-hydraulic calculations for determining time to core boiling and uncovery were performed for different plant configurations after shutdown.

Findings

The total shutdown CDF was 4.5x10–5 per reactor-year; the total full-power CDF from Seabrook's individual plant examination (IPE) was 1.1x10–4 per reactor-year.

Loss of RHR initiators contributed 82 percent to the CDF. About 71 percent of the total CDF occurred with the RCS vented and partially drained. The largest contributors to RHR failure were the hardware failure of an operating RHR pump, due to its long mission time, and the loss of RHR suction, due to either inadvertent closure of the RHR suction valves or low-level cavitation when the RCS was drained (events caused by operator error).

Although LOCAs represented only 18 percent of the total CDF, they dominated early health risks. When the RCS was filled, the equipment hatch integrity was not required (the hatch integrity is required during reduced inventory conditions). Under these conditions, a postulated LOCA would leave the operator only a short time for restoring core cooling. The Seabrook study found that it was unlikely that the equipment hatch could be closed before the containment became uninhabitable. This scenario indicated the need for controls on containment integrity and emergency response procedures for LOCA events during shutdown. This insight might have been overlooked if the level 2 analysis was not performed. A major contribution to this frequency (accounts for 8%) was LOCAs from overpressure events resulting from stuck-open RHR relief valves or ruptured RHR pump seals.

4.4 Brunswick PRA for Loss of RHR (NSAC–83)

For this study, a quantitative probabilistic evaluation was performed of the reliability of RHR equipment given a variety of scenarios in which the plant's RHR function is challenged, including following transients that resulted in reactor scrams during a planned shutdown and during a cold-shutdown scenario over time which could lead to a suppression pool temperature exceeding 200 °F (93 °C) (assumed core damage). Other functions, such as inventory control, reactivity, and containment control, were not addressed. Brunswick-specific failure data were used, and generic probability values for operational errors were included as basic events in the fault trees.

Findings

The probability of a loss of RHR during cold shutdown was estimated to be 7.0x10–6 per reactor-year. No dominant accident sequences were listed. However, it is important to note that the PRA did not include losses of inventory control which could be dominant contributors to shutdown risk.

On the basis of an evaluation of the methodology, models, and findings presented in the report, the following are major contributors to the loss of RHR during shutdown:

- RHR and RHR service water (SW) equipment unavailable due to maintenance
- RHR and RHRSW pump failures
- common-mode failure of RHR heat exchangers

4.5 Sequoyah LOCA in Cold Shutdown

Science Applications International Corporation addressed the probability of a core-melt accident in cold shutdown (Mode 5) which was initiated by a postulated loss-of-coolant accident (LOCA) at the Sequoyah nuclear plant. Two LOCA initiating events were considered: safe-shutdown earthquake and operator error (RHR-induced LOCAs were not considered). A total of 20 cases were analyzed with varying assumptions regarding time of LOCA initiation following a shutdown, LOCA size, availability of offsite power, and maintenance status.

Findings

The postulated core-melt frequency was estimated to be in the range from 7.53x10–5 to 8.5x10–7 per reactor-year. The major contributors to core-melt frequency included the following:

- operator-induced LOCAs
- availability of power to plant equipment
- maintenance
- operator errors during response (lack of procedures for securing equipment, inadequate RCS monitoring equipment)
- failure of an airbound RHR pump
- RHR suction failure

4.6 International Studies

The staff gained significant insights from studies performed in France. These studies focused on identifying the dominant contributors to risk from dilution events at shutdown and loss of RHR during midloop operation. The main PRA study excluded such external events as fires, floods, earthquakes, and source terms. The French categorized this study as a level 1 PRA.

4.7 Grand Gulf PRA for Shutdown Operation (Coarse Screening Study and Detailed Study)

Sandia National Laboratories (SNL) is performing a PRA of the low-power and shutdown modes of operation at the Grand Gulf nuclear plant for the NRC. This study has two phases. Phase 1 consisted of a screening study to determine which accident sequences need to be analyzed in more detail.* Phase 2 is the detailed analysis of the dominant accident sequences identified in Phase 1. The PRA is performed in two parts: the accident frequency analysis (level I) and the accident progression and consequence analyses (level II/III).

One objective of the screening study has been to identify plant operational states (POSs) or initiating events, or both, that require more detailed analysis during Phase 2 of the quantification process. The coarse screening in Phase 1 identified and quantified initiating events for seven POSs.** POS 5, which includes both cold shutdown and refueling modes, was selected for the detailed analysis of Phase 2. Some of the major contributing factors in selecting POS 5 over POS 4 are (1) the plant is in POS 5 for more time and (2) the technical specifications allow for more equipment to be inoperable in POS 5 during cold shutdown than in POS 4 during hot shutdown. As POS 5 was modeled in more detail, new initiating events were identified.

To simplify the development of the event trees, SNL divided the event trees into three types: (1) generic functional event trees which are at the functional level and apply to any transient, (2) generic system-level event trees which are at the mitigating systems level and form the basis for the event tree models for each specific transient initiating event, and (3) specific system level event trees which model the mitigating system's response to each of the 34 specific initiating events being analyzed in the study. SNL is currently in the process of quantifying the detailed accident sequences.

4.8 NRC Shutdown PRA for Surry (Coarse Screening Study and Detailed Analysis)

Brookhaven National Laboratory (BNL) performed a probabilistic risk assessment (PRA) of the low-power and shutdown modes of operation at the Surry nuclear plant

for the NRC. Like the Grand Gulf study discussed in Section 4.7, this study has two phases. Phase 1 consisted of a screening study to determine which accident sequences need to be analyzed in more detail. Phase 2 is the detailed analysis of the dominant accident sequences identified in Phase 1. The PRA is performed in two parts: the accident frequency analysis (level I) and the accident progression and consequence analyses (level II/III).

The objectives of Phase 2 of this program were to (1) estimate the frequencies of severe accidents that might be initiated during midloop operation, (2) compare the estimated core-damage frequencies, important accident sequences, and other qualitative and quantitative results of this study with those of accidents initiated during full-power operation, and (3) demonstrate methodologies for accident sequence analysis for plants in modes of operation other than full power.

The approach BNL used was to define different outage types and different plant operational states (POSs) within each outage type. The outage types were grouped into four types: refueling, drained maintenance, nondrained maintenance with the use of the residual heat removal (RHR) system, and nondrained maintenance without the use of the RHR system. The POSs were then used to represent the activities in the plant throughout an outage from low-power operation back to power. In a refueling outage, as many as 15 POSs were used to define the plant activities.

Three POSs defining midloop operation were selected for detailed analysis—POSs R6 and R10, occurring during a refueling outage, and POS D6, occurring during a drained maintenance outage. The detailed analysis included analyses of the initiating events, development of event trees, thermal hydraulics in support of the event tree development and accident sequence quantification, quantification of the fault and event trees using point estimates only, and assessment of human reliability.

BNL also performed a fire risk analysis for low-power and shutdown operations at Surry Unit 1. The analysis included component-based, transient-fueled, and cable fires with the frequencies for the fire events developed using the latest available information. Fire scenarios were analyzed for identified critical locations. The analysis included: (1) quantification of the initiators and the impact on the safety systems; (2) spurious operation and fire growth; (3) quantification of suppression analysis and level 1 fire risk; and (4) modification of the internal PRA model to include the impacts of fire scenarios and the scenario-dependent human reliability errors (HRPs).

Findings

The dominant contributor to the calculated core-damage frequency was the failure of the operator to mitigate the accidents. POS D6 is the most dominant POS with a core-damage frequency of 3.0×10^{-5} per reactor-year. The

---

*Phase 2 presentation to NRC staff, March 23, 1992.

**POS 4 consists of operational condition (OC) 3 with the unit on RHR/SDC, POS 5 consists of OC 4 (mode of operation) and OC 5 until the vessel head is off and level is raised to the steam lines.

characteristics of POS D6 are high decay heat levels and relatively short time available for operator action. In contrast, POS R10 has a very low decay heat and its core-damage frequency is approximately 2 orders of magnitude lower (4.7×10^{-7} per reactor-year). The overall point estimate of the core-damage frequency during low-power and shutdown operation is 3.4×10^{-5} per reactor-year.

The fire areas identified as critical are the emergency switchgear room, the normal switchgear room, the cable vault and tunnel, the containment, and the main control room. The control room fire analysis, excluding the quantification of risk, was done separately.

The quantification indicated that certain scenarios in the H and J compartments of the emergency switchgear room, one scenario in the cable vault and tunnel, and one containment scenario dominate the risk. The core-damage frequency due to fire events at midloop is estimated as 2.03×10^{-5} per reactor-year.

4.9 Findings

Quantitative results of the PRA studies are shown in Figure 4.1. *On the basis of the findings from each of the

---

*Quantitative results are not yet available from the Grand Gulf study.

studies examined above, the most significant events, from a shutdown-risk perspective, can be summarized as follows:

- failures during midloop operation (PWRs)

- operator error, especially

    - failure to determine the proper actions to restore shutdown cooling (especially during midloop)

    - procedural deficiencies

- loss of RHR shutdown cooling, especially

    - operator error induced

    - suction valve trips

    - cavitation due to overdraining of the RCS

- loss of offsite power

- LOCAs, especially

    - operator error induced

    - stuck-open RHR relief valves

    - ruptured RHR pump seals

    - temporary seals ruptured

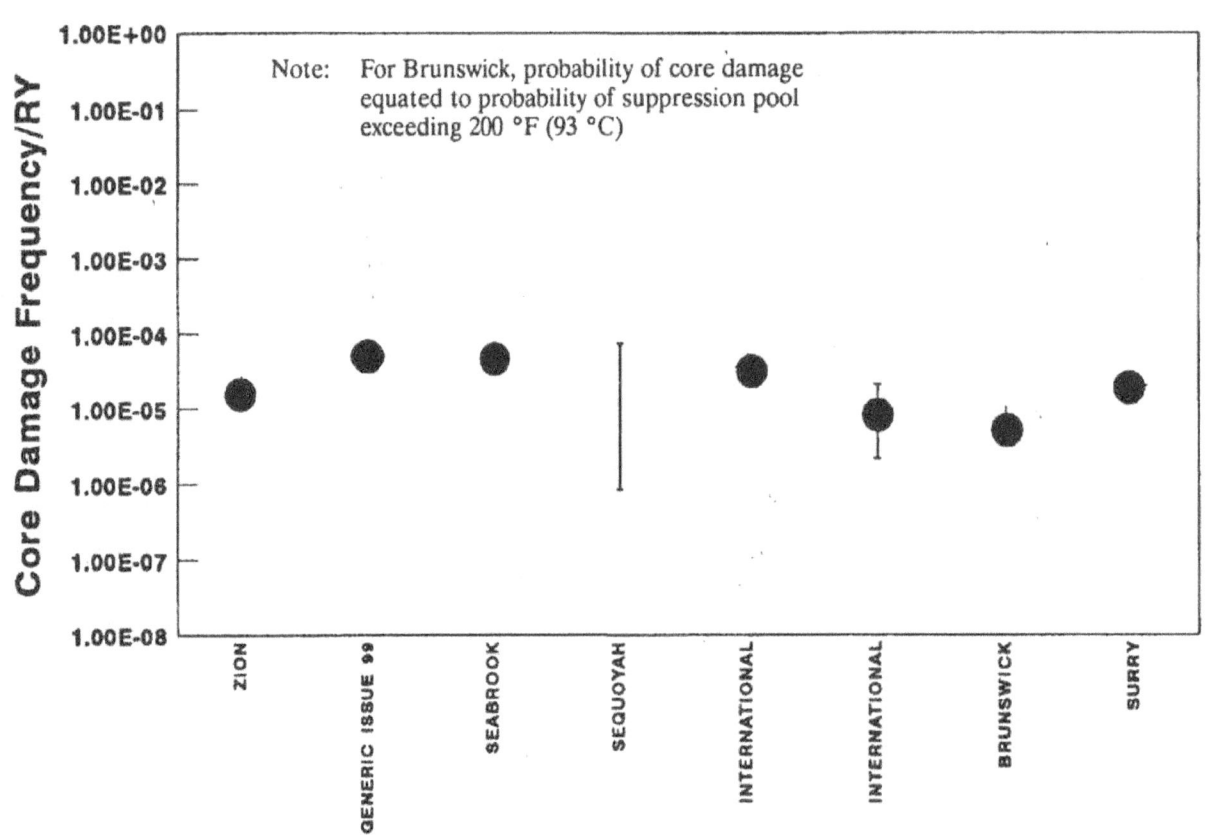

Figure 4.1 Summary of PRA results

5 REGULATORY REQUIREMENTS FOR SHUTDOWN AND LOW-POWER OPERATIONS

U.S. requirements and requirements in other countries were compiled as part of an Organization for Economic Cooperation and Development/Committee on Nuclear Regulatory Activities study led by the NRC. The findings are presented in the Nuclear Energy Agency's November 1992 proprietary report, "Regulatory Requirements and Experience Related to Low-Power and Shutdown Activities," NEA/NRA/DCOC(91)2, Revision 2, and are summarized below. No proprietary data were used.

5.1 Facilities in the United States

5.1.1 Technical Specifications

Two types of regulatory requirements address shutdown and low-power operations: design requirements and operational requirements. The regulatory design requirements contained in the general design criteria (GDCs) in Appendix A to 10 CFR Part 50 and the quality assurance requirements in Appendix B to 10 CFR Part 50 do not generally depend on operational mode. The staff has interpreted the GDC requirements in the regulatory guides and the "Standard Review Plan," NUREG-0800.

The technical specifications for individual plants are the primary sources of operational requirements to control shutdown and low-power operation. The current standard technical specifications (STS) address specific requirements during shutdown and low-power operation for reactivity control, inventory control, residual heat removal, and containment integrity. The STS requirements vary in degree of coverage and allowable limits when compared with those issued earlier in custom technical specifications. The following discussions of technical specifications are based on the current STS for pressurized-water reactors (PWRs) and boiling-water reactors (BWR/4s).

5.1.1.1 Reactivity Control

The technical specifications requirements for PWRs during shutdown operation include a reduction in the shutdown margin from 1.6-percent to 1.0-percent delta K/K during cold shutdown. Reactor protection systems are not required to be operable once the reactor is shut down, except that flux monitors must be operable whenever control rods can be moved. The restoration of an inactive loop is controlled by temperature and boron concentration limits during cold shutdown and refueling. Boron concentration limits are not applicable for the refueling water storage tank (RWST) during hot and cold shutdown and refueling operations, and the boron injection tank is not required to be operable during cold shutdown and refueling. However, sources of unborated water must be isolated from the primary system.

For BWRs, operability requirements of relevant components of the reactor protection system remain in effect in the low-power and shutdown/refueling "Operation Conditions." These include operability of the SRM (source range), IRM (intermediate range), APRM (average power range) flux monitors (and trip systems), control rod scram accumulators, scram discharge volume, mode switch, and manual scrams. There are STS requirements on control rod insertion (for the most part minimizing withdrawn rods) and on shutdown margin. These are augmented by special STS for refueling operations and for special tests exceptions. For example, when permitted, multiple control rod removal requires prior removal of all fuel in the affected control cell. However, if control rods are being moved, flux monitors must be operable. The feedwater reactor trip may be disabled during the startup mode and the instrumentation for anticipated transient without scram (ATWS) is not required during startup. All control rod movement is restricted to one control blade at a time, unless the associated fuel cell contains no fuel. The shutdown margin must be at least 0.38-percent delta K/K at all times.

5.1.1.2 Inventory Control

For both PWRs and BWRs, leakage limits and leakage detection system operability are not required during cold shutdown and refueling. The following additional requirements apply only to PWRs: Only one train of emergency coolant injection is required during hot shutdown and none is required in cold shutdown or refueling. The RWST is also not required to be operable during cold shutdown or refueling. Instrumentation requirements are controlled by the requirements of the systems supported by the instrumentation; that is, if the injection system is required to be operable, the system instrumentation is required to be operable. In addition, for PWRs, low-temperature overpressure protection is required in the hot-shutdown, cold-shutdown and refueling conditions. The requirements are that two power-operated relief valves or two residual heat removal (RHR) relief valves are operable and no more than one train of high-pressure injection can be operable.

For BWRs, two low-pressure injection trains are required during cold shutdown and refueling. This requirement is eliminated if the reactor pressure vessel (RPV) head is removed, the refueling cavity is flooded, spent fuel pool gates are removed, and the level is maintained as required by technical specifications (TS). As with the PWR instrumentation requirements, the system instrumentation is required to be operable if the system is required to be operable. Cooling water systems associated with the injection systems are also generally required to be operable only when the injection systems are required to be operable, unless required to meet other TS requirements.

5.1.1.3 Residual Heat Removal

In the low-power and shutdown modes, the PWR operability requirements for the RHR function are mode dependent. During hot standby, two reactor coolant loops are required. In hot shutdown, any combination of two RHR loops and reactor coolant loops is acceptable. During cold shutdown, two RHR loops are required, unless two steam generators are filled to at least 17 percent of the normal level for the steam generators; in that case, two steam generators and one RHR loop are an acceptable combination. During refueling, two RHR loops or one with the refueling cavity filled are required. Generally, the secondary-side heat removal systems (main and auxiliary feedwater) are not required to be operable during hot and cold shutdown and refueling. However, if a steam generator is being used as a heat removal system during hot shutdown, the condensate storage tank, atmospheric dump valves, and one train of auxiliary feedwater (including instrumentation) must be available.

For BWRs, two loops of RHR are required (with one operating) in the hot-shutdown, cold-shutdown, and refueling modes. With the refueling cavity flooded during refueling, only one RHR loop is required.

One division of electric power is required to be operable in cold shutdown and during refueling, as opposed to two divisions during all other modes of operation. (A division is defined to include both an onsite and an offsite source of ac power.)

5.1.1.4 Containment Integrity

The containment integrity requirements for PWRs are not applicable during cold shutdown and refueling. This includes the operability of the containment spray system. In addition, the containment isolation instrumentation is not required to be operable during hot shutdown. During fuel movement operations, less-restrictive containment isolation requirements are in effect. One airlock door must be maintained closed and a "four-bolt rule" is in effect for the equipment hatch.

In a BWR, the containment atmosphere can be de-inerted 24 hours prior to being at a power level less than 15 percent of rated thermal power. The primary containment must be inerted within 24 hours after exceeding 15 percent of rated thermal power during startup. Primary containment integrity and containment isolation instrumentation requirements are not applicable during cold shutdown and refueling. However, during fuel movement, core alterations, and operations with the potential

for draining the vessel, both the secondary containment and the standby gas treatment system must be operable.

The staff is reviewing the range of TS requirements for shutdown and low-power modes, including those in the existing STS and those developed within the Technical Specifications Improvement Program. In performing this review, the staff has determined that these requirements are generally less restrictive than the requirements in the full-power operations mode. For example, the TS allow fewer operators for PWRs and BWRs during cold-shutdown and refueling operations.

5.1.2 Other Regulatory Requirements or Policies

The staff also identified a number of important facts regarding regulatory requirements or policies pertaining to operator training, use of overtime, emergency planning, fuel handling, heavy loads, fire protection, and procedures.

5.1.2.1 Training (Coverage of Shutdown Conditions on Simulators)

The current Code of Federal Regulations (Title 10, Section 55.45(b)(2)(iv)) requires that the simulation facility portion of the operating test only be administered on a certified or approved simulation facility. NRC Regulatory Guide 1.149 endorsed the guidance of the American National Standards Institute's (ANSI's)/American Nuclear Society's (ANS's) standard, "Nuclear Power Plant Simulators for Use in Operator Training," ANSI/ANS 3.5–1985. To date, nearly all of the industry's simulators have been certified to meet this guidance.

The ANSI/ANS Standard 3.5–1985 requires simulation of minimum normal activities from cold startup to full power to cold shutdown, excluding operations with the reactor vessel head removed.

5.1.2.2 Policy on Use of Overtime

Generic Letter (GL) 82–12 transmitted NRC's "Policy on Factors Causing Fatigue of Operating Personnel at Nuclear Power Plants." This policy gives specific guidance for the control of working hours during shutdown operations. This guidance allows the plant superintendent to approve associated deviations from the guidelines on working hours. The policy applies only to personnel who perform safety-related duties and the individuals who directly supervise them.

5.1.2.3 Fire Protection

The plant TS allow various safety systems, including fire protection systems, to be taken out of service to facilitate system maintenance, inspection, and testing during shutdown and refueling.

The Appendix R (10 CFR Part 50) fire protection criteria for protecting the safe-shutdown capability does not include those systems important to ensuring an adequate level of RHR during non-power modes of operation.

The current Nuclear Regulatory Commission (NRC) fire protection philosophy (NUREG-0800, Standard Review Plan Section 9.5.1) does not address shutdown and refueling conditions and the impact a fire may have on the plant's ability to remove decay heat and maintain reactor coolant temperature below saturation conditions.

5.1.2.4 Reporting Requirements

The current NRC regulations require that any operation or condition prohibited by the plant TS is reportable under 10 CFR 50.73. This includes both power operation and shutdown. However, as discussed earlier, there are far fewer TS applicable during shutdown.

5.1.2.5 Onsite Emergency Planning

The current guidance for classifying emergencies for nuclear plants during power operation (found in Appendix I to NUREG-0654 (FEMA–REP–1), Revision 1, titled "Criteria for Preparation and Evaluation of Radiological Emergency Response Plans and Preparedness in Support of Nuclear Power Plants"), does not explicitly address the different modes of nuclear power plant operation.

5.1.2.6 Fuel Handling and Heavy Loads

The following requirements apply to the handling of fuel:

(1) Plant TS require that fuel-handling equipment be tested before use, in order to prevent dropping fuel elements.

(2) For both BWRs and PWRs, TS require that a specified level of water be maintained above the reactor vessel head and in the spent fuel storage pools during refueling, in order to maintain spent fuel cooling capability and to ensure protection against radiation in the event of a fuel-handling accident.

(3) For PWRs, TS require that, before initiation of the refueling process, penetrations in the containment building be closed or be capable of being closed by an operable automatic valve actuated on a high-radiation signal in the containment. For BWRs, TS require that the integrity of the secondary containment be ensured before handling irradiated fuel. The reason for maintaining containment integrity in both PWR and BWR plants is to prevent excessive radiation from entering the environment in the event of an accident.

(4) For PWRs and BWRs, commitments in final safety analysis reports (FSARs) and license amendments require that specified coolant temperatures be maintained in the spent-fuel pool and that fuel pool cooling systems be operating to maintain those temperatures. TS require that specified water levels be maintained in the spent-fuel pool.

(5) TS require that, before initiating any core changes by means of fuel handling, the reactor be subcritical for a specified period, in order to permit decay of short-term fission products.

(6) TS require that shutdown margins be ensured before starting the refueling process in order to prevent plant criticality during that period.

(7) TS ensures operation and operability of core cooling systems are ensured before and during the refueling process.

(8) As an added protection against excessive radiation escaping to the environment, PWR TS require that containment purge and exhaust isolation systems be operable to isolate the primary containment in the event of a fuel-handling accident. TS for PWRs also require that storage pool cleanup systems be operable to filter and remove radioactive material from the atmosphere over the spent-fuel pool should an accident occur while fuel is being moved. For BWRs, TS require that secondary containment integrity be ensured (as noted above) and that the standby gas treatment system be operable during the refueling procedure.

Heavy Loads

Damage to spent fuel or redundant trains of safe-shutdown systems is prevented by following safe load paths, i.e., by circumventing of areas in which spent fuel is stored or where safe-shutdown systems are located. Where a licensee cannot employ safe load paths at all times, protection is afforded by one of two alternative methods:

(1) by providing a single-failure-proof handling system, or

(2) by completing an acceptable analysis of a potential drop of a heavy load, as follows:

(a) For shutdown systems, heavy load drops are analyzed to ensure continued operability of at least one train of redundant systems.

(b) For spent fuel, heavy load drops are analyzed: (i) to ensure protection against escape of radiation; (ii) to ensure protection against criticality of spent fuel arrays, and (iii) to ensure protection against damage to reactor vessels and spent-fuel pools which would prevent continued cooling capability of spent fuel.

5.1.2.7 Plant Procedures

Appendix B to 10 CFR Part 50 requires that licensees provide control over activities affecting the quality of plant structures, systems, and components that prevent or mitigate the consequences of postulated accidents that could cause undue risk to the health and safety of the public. The control of these structures, systems, and components is to be consistent with their importance to safety, and includes maintaining safety during shutdown as well as power operation. Activities affecting quality are to be performed in accordance with procedures or drawings of a type appropriate to the circumstances. Consequently, the regulatory basis now exists to require that licensees have procedures appropriate for the prevention and mitigation of risks associated with low-power and shutdown operations and to require that these procedures are commensurate with the risk to public health and safety.

5.1.3 Bulletins and Generic Letters

NRC use of generic communications, specifically bulletins and generic letters, offers insight into the events of interest and the evolution of requirements. These generic communications present a chronology of events and actions requested by the NRC (actions for plant licensees to take to preclude or mitigate events that could affect the nuclear power plant during low-power and shutdown operations) that have resulted in changes to regulatory requirements.

Two generic letters (87-12 and 88-17) are of interest to low-power and shutdown operations. They contain actions requested of licensees or identify actions taken by licensees. They are the most comprehensive and most widely applicable of the generic letters. They specifically address shutdown concerns and are the most current generic letters to contain recommendations regarding low-power and shutdown operations.

Table 5.1 lists seven generic letters related to shutdown and low-power operations and Table 5.2 lists the requirements and recommendations of GL 88-17.

Bulletin 80-12 is related to shutdown and low-power operations. It discusses the loss of decay heat removal that occurred at Davis-Besse in April 1980 and delineates the actions to be taken by the industry.

5.2 International Facilities

In January 1991, the Committee on Nuclear Regulatory Activities (CNRA) sent a questionnaire to the regulatory agencies of several nations. This questionnaire, "Elements for a Survey on Low-Power and Shutdown Activities," was intended to gather information regarding approaches to the control of low-power and shutdown operations at nuclear power plants. The objective of the questionnaire was that the responses would address all low-power and shutdown requirements, both of the regulatory authority and of the facility operators. However, most responses addressed the regulatory requirements and simply acknowledged that operation during these modes was mainly controlled by procedures and requirements established by the facility operator.

In particular, the responses were to address requirements for reactivity control, inventory control, residual heat removal, containment integrity, and outage and maintenance management. Each country indicated that its regulatory body has established safety requirements that the operator had to meet. However, the specific operating requirements were developed by the plant operator.

Technical specifications or their equivalent appeared to be the principal technique used to impose regulatory control of plant operation during shutdown and low-power operation.

Table 5.1 Generic Letters Concerning Shutdown and Low-Power Operations

Generic Letter	Title
80-53	Transmittal of Revised Technical Specifications for Decay Heat Removal Systems at PWRs
81-21	Natural Circulation Cooldown
85-05	Inadvertent Boron Dilution Events
86-09	Technical Resolution of Generic Issue B-59, (n-1) Loop Operation in BWRs and PWRs
87-12	Loss of Residual Heat Removal (RHR) While the Reactor Coolant System (RCS) Is Partially Filled
88-17	Loss of Decay Heat Removal
90-06	Resolution of Generic Issues 70, "Power-Operated Relief Valve and Block Valve Reliability," and 94 "Additional Low-Temperature Overpressure Protection for Pressurized Water Reactors" [pursuant to 10 CFR 50.54(f)]

Table 5.2 Generic Letter 88–17* Recommendations and Program Enhancements

Item	Recommendation
(1)+	Discuss with appropriate plant personnel the Diablo Canyon event, lessons learned, and implications. Provide training shortly before entering a reduced inventory condition.
(2)+	Implement procedures and administrative controls that reasonably ensure containment closure will be achieved before the time at which a core uncovery could result from a loss of decay heat removal coupled with an inability to initiate alternate cooling or to add water to the reactor coolant system.
(3)+	Provide at least two independent, continuous temperature indications that are representative of the core exit conditions whenever the reactor is in midloop operation and the reactor vessel head is located on top of the vessel.
(4)+	Provide at least two independent, continuous reactor coolant system water level indications whenever the reactor coolant system is in a reduced inventory condition.
(5)+	Implement procedures and administrative controls that generally avoid operations that deliberately or knowingly lead to perturbations to the RCS or to systems that are necessary to maintain the RCS in a stable and controlled condition while the RCS is in a reduced inventory condition.
(6)+	Provide at least two available or operable means of adding inventory to the reactor coolant system in addition to the pumps that are a part of the normal decay heat removal systems.
(7)+	Implement procedures and administrative controls that reasonably ensure that all hot legs are not blocked simultaneously by nozzle dams unless a vent path is provided that is large enough to prevent pressurization of the upper plenum of the reactor vessel.
(8)+	Implement procedures and administrative controls that reasonably ensure that all hot legs are not blocked simultaneously by closed loop stop valves unless reactor vessel pressurization can be prevented or mitigated.
(9)#	Provide reliable indication of parameters that describe the state of the reactor coolant system and the performance of systems normally used to cool the reactor coolant system for both normal and accident conditions. The following should be provided in the control room: two independent indications of reactor vessel level and temperature, indications of decay heat removal system performance, and visible and audible indications of abnormal conditions.
(10)#	Develop and implement procedures that cover reduced inventory operation and that provide an adequate basis for entry into a reduced inventory condition.
(11)#	Ensure that adequate operating, operable, or available equipment is provided for cooling the reactor coolant system. Maintain existing equipment in an operable or available status, including at least one high-pressure system and one other system. Provide adequate equipment for personnel communications.
(12)#	Conduct analyses to supplement existing information and develop a basis for procedures, instrumentation installation and response, and equipment/nuclear steam supply system interactions and response.
(13)#	Identify technical specifications that restrict or limit the safety benefit of these actions and submit appropriate changes.
(14)#	Reexamine recommending item 5 and refine it as needed.

*This generic letter discussed the loss of decay heat removal capability that occurred on April 10, 1987, at Diablo Canyon Unit 2 while the plant was in the refueling mode of operation. Additional events at Waterford (on May 12, 1988), Sequoyah (on May 23, 1988), and San Onofre (on July 7, 1988) also contributed to this second generic letter addressing loss of decay heat removal capabilities at PWRs. It provided recommendations and required PWR licensees to respond to the recommendations.

+ Recommended for implementation before operating in a reduced inventory condition.

Recommended for implementation as soon as practical.

These requirements were generally less restrictive in the shutdown mode than in the full-power operations mode. Low-power operation was often approached with the same requirements as full-power operation, although in specific instances the technical specifications requirements during low power were relaxed from the full-power requirements.

Of the areas addressed in the questionnaire, the outage and maintenance management area appeared to be the most within control of the operators of the nuclear facility. General requirements to submit outage plans and refueling documentation were the most restrictive of the requirements imposed by any country, and most appeared to require some type of planning. In the other areas addressed by the questionnaire, some control over the plant configuration was exercised in the technical specifications (or their equivalent) in most countries.

Reactivity control requirements for PWRs tended to address two related items: boron concentration (including both boron injection system operability and the need to isolate the primary system from sources of non-borated water) and subcriticality margin. Additional requirements mentioned in many responses included requirements to maintain neutron flux monitoring instrumentation operable in all modes, unless the control rods cannot be moved.

Generally, fewer reactivity control requirements were imposed on the BWRs than on PWRs. During refueling operations, restrictions were generally in place regarding the removal of control assemblies from the core. Either one rod at a time was allowed to be removed or the supercell around the control rod to be removed must be empty.

Several different approaches were taken to describe the inventory control requirements. Some countries described the instrumentation requirements for the shutdown and low-power operational modes. For these countries, additional instrumentation was required at various times during operation in these modes, particularly during PWR midloop operations.

The responses from several countries described injection capability requirements. Combinations of low- and high-pressure injection systems were required to be operable. Often, during the time that the refueling cavity was flooded, the injection system requirements were reduced. However, if maintenance was being performed on the primary system below the level of the core, this reduction in injection availability was not allowed.

In general, redundant heat-removal capabilities were required at all times by most of the countries. In PWRs, this redundancy could often be supplied by any combination of operable steam generators and RHR systems, shifting entirely to the RHR systems once the steam generators cannot be used. For those countries that replied in detail, their responses indicated that the flooded refueling cavity can be considered a heat-removal system, because of the large amount of water present. At least two countries tied the operability of the RHR system to the decay heat rate as a function of time after shutdown. For these countries, the requirements on system operability were reduced as the decay heat rate dropped.

In general, containment integrity requirements were waived under certain conditions in every country. Usually, during the refueling mode of operation when no fuel transfer was taking place, containment integrity was not required. Containment airlocks were not always required to remain operable during refueling. When they were allowed to be open during refueling, they must generally isolate on a high radiation signal. In BWRs with inerted containments, the containment generally may be de-inerted several hours before entering a cold-shutdown condition and did not have to be re-inerted until after entering a hot-shutdown condition.

Other than some staffing requirements, there were almost no regulatory requirements that specifically addressed outage and maintenance management. Many countries did require that outage and refueling plans be submitted to the regulatory bodies. These documents must outline the procedures and rules to be followed during an outage. However, the licensee generally developed the procedures and rules.

Significant variability appears to exist among the programs in various countries.

Conclusions

The NRC's current requirements in the areas of shutdown and low-power operations were less stringent than those of most other regulatory agencies. However, the staff concluded that the NRC's continuing shutdown-risk study appears to address all the significant issues.

6 TECHNICAL FINDINGS AND CONCLUSIONS

6.1 Overview

On the basis of the work it completed over the past 18 months, the staff concludes that risk varies widely during shutdown conditions at a given plant and among plants, and can be significant. The staff has observed an increasing recognition of the importance of shutdown issues among licensees and within the staff. The staff also observed a general improvement in safety practices during shutdown, both in response to regulatory actions and from the industry's individual and collective initiatives.

Variability of risk during an outage period results primarily from continuous changes in (1) plant configuration and activity level, which determine the likelihood of an upset and, to some degree, the severity; (2) the amount and quality of equipment available to recover from an upset; (3) the time available to diagnose and recover from an upset; and (4) the status of the primary containment. Among plants, risk varied because of the many approaches used by utilities to address safety during a shutdown condition, differences in plant design features, and lack of a standard set of industry or regulatory controls for shutdown operations. Such variability, along with analytical limitations peculiar to shutdown (e.g., human reliability analysis), makes it difficult to quantify the risk during shutdown in U.S. reactors. The staff has focused its attention primarily on operating experience and the current capability in U.S. plants to avoid a core-melt accident and release of radioactivity. Insights from probabilistic assessments have also been valuable in understanding what is important to risk during shutdown.

As discussed in Chapter 1, about midway through the evaluation the staff identified a number of issues believed to be especially important and a number of potentially important issues. The staff has studied each of these issues and obtained specific findings which are discussed in this chapter.

6.2 Outage Planning and Control

In the absence of strict technical specification controls, licensees have considerable freedom in planning outage activities. Outage planning determines what equipment will be available and when. It determines what maintenance activities will be undertaken and when. It effectively establishes if and when a licensee will enter circumstances likely to challenge safety functions and it establishes the level of mitigation equipment available to deal with such a challenge.

Many shutdown events have occurred that represented challenges to safety during low-power and shutdown (LPS) operation. Some of these initiated when the power plant was in a sensitive condition as a result of inadequate planning and mistakes (examples: Diablo Canyon, 4/87, see NUREG-1269; Vogtle, 3/90, see NUREG-1410). Recognizing that the safety significance of such events is a strong function of outage planning and control, and that the NRC has not previously addressed the safety implications of outage planning, the staff initiated a study of such planning and its implications as part of the plant visits program described in Chapter 3, and has supplemented this with information from staff inspectors.

A wide variety of conditions and planning approaches was observed during the plant visits. These included

* outages that were well planned and controlled

* outages that were poorly prepared and poorly organized

* priority assigned to safety with the complete licensee organization striving for safety

* an ad hoc approach in which safety was dependent upon individual judgment

* the perception that short outages represent excellence

* personnel stress and events that appeared to be the result of overemphasis on achieving a short outage

* impact of poor outage planning and implementationon on plant operation

* imprudent operation as a result of insufficient attention to safety

6.2.1 Industry Actions

The industry has addressed outage planning and control with programs that include workshops, Institute of Nuclear Power Operations (INPO) inspections, Electric Power Research Institute (EPRI) support, training, procedures, and other programs. One activity (a formal initiative proposed by the Nuclear Management and Resources Council (NUMARC) has produced a set of guidelines for utility self-assessment of shutdown operations (NUMARC 91-06); these guidelines serve as the basis for an industrywide program that was implemented at all plants by December 1992. This provides high-level guidance that addresses many outage weaknesses. Detailed guidance on developing an outage planning program is beyond the scope of the NUMARC effort.

NUMARC 91-06 states: "The underlying premise of this guidance is that proper outage planning and control, with a full understanding of the major vulnerabilities that are present during shutdown conditions, is the most effective means of enhancing safety during shutdown."

The staff met with NUMARC and the associated utility working group on several occasions to share technical insights and discuss program status. The initiative does appear to be a significant and constructive step and effects may have already been realized by a few utilities using draft guidance in recent outages.

6.2.2 NRC Staff Findings

On the basis of its review of operating experience, probabilistic risk assessments (PRAs), site visits, and information from other regulatory agencies, the staff concludes that a well-planned, well-reviewed, and well-implemented outage is a major contributor to safety. It has further substantiated and/or determined the following:

- Consistent industrywide safety criteria for the conduct of LPS operation do not exist. (NUMARC 91-06 provides high-level guidance, but no criteria.)

- Many licensees have no written policy that provides safety criteria for LPS operation. Some are working on such a policy; others had no plan (at the time the staff visited the plant) to prepare such a policy.

- Some licensees enter planned outages with incomplete outage plans.

- Some licensees cannot properly respond to an unscheduled outage because their planning is poor.

- Safety considerations are not always evident during outage planning.

- Changes to outage plans and ad hoc strategies for activities not addressed in the plan are often not addressed as carefully as the original plan.

- The need for training and procedures is not always well addressed in planning.

- Bases do not exist that fully establish an understanding of plant behavior and that substantiate the techniques depended upon to respond to events. Such bases would provide the information necessary for reasonable and practical technical specifications, procedures, training, LPS operation (outage) planning, and related topics.

- There is no regulation, regulatory basis, staff policy, or other guidance (such as technical specifications or staff studies) that currently requires or otherwise

provides regulatory guidance for outage planning and plan implementation.

6.3 Stress on Personnel and Programs

A large amount of activity takes place during outages. The increased size of the work force at the site during outages, combined with the rapid changes in plant configurations that occur during these periods, creates a complex environment for planning, coordinating, and implementing tasks and emergency responses. As a result, outage activities can stress the capabilities of plant personnel and programs responsible for maintaining quality and operational safety. This stress can be reduced through outage planning that ensures (1) staffing levels are sufficient and jobs are defined so that workloads during normal or emergency outage operations do not exceed the capabilities of plant personnel or programs; (2) personnel are adequately trained to perform their duties, including the implementation of contingency plans; and (3) contingency plans are developed for mitigating the consequences of events during shutdown.

The present NRC policy concerning working hours of nuclear plant staff, as written, provides objectives for controlling the working hours of plant personnel, and provides specific guidelines for periods when a plant is shut down. It permits plant personnel to work overtime hours in excess of the recommended hours, provided that appropriate plant management gives its approval. However, as noted in NRC Information Notice 91-36, in some instances a licensee's work-scheduling practices or policies were inconsistent with the intent of the NRC policy.

The staff reviewed the NUMARC document "Guidelines to Enhance Safety During Shutdown" and concluded that the guidelines establish a sound approach to addressing the issue of stress and its risks associated with LPS operations. Effective implementation of these guidelines should reduce the potential for planned or unplanned outage activities to inappropriately stress the capabilities of plant personnel and programs by (1) improving control of outage activities, (2) reducing time that people perform higher risk activities, and (3) increasing preparedness to implement contingency actions, if needed. Consequently, stress on plant programs and personnel during outages is expected to be reduced.

6.4 Operator Training

Conditions and plant configurations during shutdown for refueling can place control room operators in an unfamiliar situation. Operators who are properly informed and who understand the problems that could arise during outages are essential in reducing risks associated with the outage activities. Through the comprehensive training programs, operators can gain such knowledge and understanding, thus increasing the level of safe operations at nuclear plants. The level of knowledge and abilities can be qualitatively measured by a comprehensive examination.

6.4.1 Examination of Reactor Operators

The knowledge and abilities (K/A) that an operator needs to properly mitigate the events and conditions described in Chapters 2 and 3 are addressed by NRC's K/A catalogs (NUREG-1122 and NUREG-1123). These catalogs, in conjunction with the facility licensee's job task analysis, provide the basis for developing examinations that contain valid content. Present guidance for developing examinations is described in the Examiner Standards (NUREG-1021). This guidance allows for significant coverage of shutdown operations, but it does not specify any minimum coverage. NUREG-1021 provides a methodology for developing examinations that was derived, in part, from data collected from licensed senior reactor operators and NRC examiners. The guidance also calls for examination content to include questions and actions based on operating events at the specific facility and other similar plants. A review of samples of initial written examinations indicates that LPS operations are covered generally and the coverage is consistent with assuring adherence to the objectives of licensee training programs and the sampling methodology of NUREG-1021. However, if licensee training programs and procedures are revised, through an improved outage program, to place more emphasis on reducing shutdown risks, the staff expects that more extensive and broader examination coverage will follow.

6.4.2 Training on Simulators

As of May 26, 1991, all facility licensees were required to have certified or approved simulation facilities unless specifically exempted. Nearly all of the industry's simulators have been certified to meet the guidance of the American National Standards Institute (ANSI) "Nuclear Power Plant Simulators for Use in Operator Training," ANSI/ANS 3.5-1985, as endorsed by Regulatory Guide 1.149. This standard calls for simulation of minimum normal activities from cold startup to full power to cold shutdown, excluding operations with the reactor vessel head removed. Therefore, these certified simulators are capable of performing many of the operations from a subcritical state to synchronization with the electrical grid.

ANSI/ANS 3.5-1985 is based on the concept that the scope of simulation should be commensurate with operator training needs. In accordance with ANSI/ANS 3.5-1985, the scope of simulation should be based on a systematic process for designing performance-based operator training, and modifications should be based on assessments of the training value this process offers. The

scope of the necessary changes would be defined by operator tasks identified as requiring training or examination on a simulator. Presently, simulators are used in training and examinations in those areas where dynamic plant response provides the most appropriate means to meet the training objectives. Many events that are likely to occur during shutdown would result in the majority of operator actions taking place out in the plant rather than in the control room. As a result, such events might be more appropriately addressed through methods other than simulator training.

To the extent practicable, simulator training for shutdown conditions should continue to be conducted. The Examiner Standards document (NUREG-1021) already requires examiners to report observations of simulator performance in the examination reports. This feedback from the examiners is then used to determine if simulator inspections are necessary. Revising NUREG-1021 to place more emphasis on reducing shutdown risks should result in more observations of simulator performance in this area being reported than at present.

6.5 Technical Specifications

6.5.1 Residual Heat Removal Technical Specifications

Based primarily on the PRA studies discussed in Chapter 4 and the thermal-hydraulic analysis in Section 6.6, the staff concludes that current standard technical specifications (STS) for pressurized-water reactors (PWRs) are not detailed enough to address the number and risk significance of reactor coolant system configurations used during cold shutdown and refueling operations. This is particularly true of PWR technical specifications. Safety margin during these modes of operation is significantly influenced by the time it takes to uncover the core following an extended loss of residual heat removal (RHR). The conditions affecting this margin significantly include decay heat level, initial reactor vessel water level, the status of the reactor vessel head (i.e., bolted on or bolted on with bolts detensioned or removed), the number and size of openings in the cold legs, the existence of hot-leg vents, whether or not there are temporary seals in the reactor coolant system (RCS) which could leak if the system is pressurized, and availability of diverse, alternate methods of RHR in case of complete loss of RHR systems. The current technical specifications do not reflect these observations. The staff has also found that some older plants do not have even basic technical specifications covering the RHR system.

In light of the above findings, the staff has identified a number of proposed improvements to limiting conditions for operation in current standard technical specifications for the RHR systems, component cooling water systems, service water systems, and emergency core cooling systems. These improvements are discussed in Chapter 7.

6.5.2 Electrical Power Systems Technical Specifications

Electric power and its distribution system is generally as vital for accident mitigation during shutdown conditions as it is for power operating conditions. There are, however, some shutdown conditions for which it is not as vital and during which losses of power can be accommodated more easily (e.g., fuel offload and reactor cavity flooded). In PWRs, all normal RHR systems and most components used in alternate methods are powered electrically. The same holds true for the emergency core cooling system (ECCS) and instrumentation. Boiling-water reactors (BWRs) are similar, but many more systems that are powered by steam are available to remove heat; however, these systems can only be used when the reactor vessel head is on and the main steam system is pressurized. Electric power is also vital for securing primary containment integrity promptly at some plants (see Appendix B).

Current STS were written under the assumption that all shutdown conditions were of less risk than power operating conditions. As a result of making that assumption, most maintenance on electrical systems is done during shutdown. Consequently, requirements for operability of systems are relaxed during shutdown modes.

Operating experience and risk assessments discussed in Chapters 2 and 3 indicate that for some shutdown conditions (e.g., midloop operation) such relaxation of operability requirements for electrical systems is not justified. In addition, in the past, STS in the electrical system area have been poorly integrated with technical specifications for other systems that the electrical systems must support. As a result, many plant-specific technical specifications for shutdown conditions are also poorly integrated; and misunderstandings have occurred regarding how the electrical specifications should be applied to support other technical specifications for systems such as RHR systems. There are also some facilities that do not have any electrical system technical specifications for shutdown modes.

In light of these findings and knowledge of shutdown operations gained from the site visits, the staff concludes at this time that with proper planning, maintenance on electrical systems can be accommodated during shutdown conditions of less risk significance. Consequently, the staff is developing proposed improvements to technical specifications for electrical systems which (1) ensure a minimum level of electrical system availability in all plants, (2) balance the need for higher availability of electrical systems during some shutdown conditions and the need to still do maintenance during shutdown operations, and (3) bring logic and consistency to an area of nuclear

plant operation that has been cumbersome for both plant operators and regulators.

6.5.3 PWR Containment Technical Specifications

As discussed in Chapter 5, containment integrity for PWRs and BWRs is not required by technical specifications during cold shutdown or refueling conditions, except during movement of fuel. On the basis of operating experience, thermal-hydraulic analyses, and PRA assessments, the staff concludes that it may be necessary to ensure PWR containment integrity prior to an interruption in core cooling under some shutdown conditions (this is discussed more fully in Section 6.9.1). Changing the technical specification on containment integrity would be the most direct and effective means of improving containment capability where needed. However, the staff recognizes the importance of containment access during outages and accepts that having some passive cooling methods available, in addition to normal cooling systems, can compensate for an open containment when decay heat is high. Consequently, the staff is considering the need for a proposed technical specification to govern containment integrity for PWRs during some shutdown conditions; the proposed technical specification recognizes the importance of passive alternate cooling methods, as discussed in Chapter 7.

6.6 Residual Heat Removal Capability

6.6.1 Pressurized-Water Reactors

Decay heat is removed in PWRs during startup and shutdown by dumping steam to the main condenser or to the atmosphere and restoring inventory in the steam generators with the auxiliary feedwater (AFW) system. During cold shutdown and refueling, the RHR system is used to remove decay heat. Because of the relatively high reliability of the AFW system and the short time spent in the startup and shutdown transition modes, losses of decay heat removal during these modes have been infrequent. However, loss of decay heat removal during shutdown and refueling has been a continuing problem. In 1980, a loss-of-RHR event occurred at the Davis-Besse plant when one RHR pump failed and the second pump was out of service. Following its review of the event and the requirements that existed at the time, the NRC issued Bulletin 80-12, followed by Generic Letter (GL) 80-53 calling for new technical specifications to ensure that one RHR system is operating and a second is available (i.e., operable) for most shutdown conditions. The Diablo Canyon event of April 10, 1987, highlighted the fact that midloop operation was a particularly sensitive condition. Following its review of the event, the staff issued GL 88-17, recommending that licensees address numerous generic deficiencies to improve the reliability of the decay heat removal capability. More recently, the incident investigation team's report of the loss of ac power at the Vogtle plant (NUREG-1410) raised the issue of coping with a loss of RHR during an extended period without any ac power. In light of the continued occurrence of events involving loss of RHR and the issues raised in NUREG-1410, the staff assessed the effectiveness of GL 88-17 actions and alternate methods of decay heat removal. These assessments are discussed next.

6.6.1.1 Effectiveness of GL 88-17 Actions

Actions requested in GL 88-17 are listed in Table 5.2. The staff assessed the response to GL 88-17 through NRC inspections conducted to date and the site visits discussed in Chapter 3. The more important subject areas were evaluated in terms of overall performance since GL 88-17 was issued, as discussed below.

Operations. Operations with the RCS water level at midloop have diminished generally. Some utilities now perform activities requiring reduced inventory with the reactor defueled. Others have taken steps to minimize time spent in reduced inventory or plan sensitive activities later in the outage when the decay heat level is lower. However, midloop operation is still used widely; in fact, one utility stayed at midloop for 37 days in its most recent outage.

Events. Loss-of-RHR events have continued to occur even 3 years after the issuance of GL 88-17. Three events discussed in Chapter 2 occurred in 1991. All three occurred at sites that had also experienced such events before GL 88-17 was issued.

Procedures. As discussed in Chapter 2, procedures for responding to loss-of-RHR events have generally improved in terms of the level of information provided to operators and the specification of alternate systems and methods that can be used for recovery. In addition, inspection teams have found that procedures written in response to GL 88-17 have been applied effectively outside the intended envelope for lack of other procedures, for example, loss of inventory.

However, some concerns still exist. Although procedures often specify use of the steam generators or the ECCS as alternate methods for removing decay heat, it has been observed, as discussed in Chapter 3, that neither steam generator availability nor a clear flow path via the containment sump has been planned for and maintained. In addition, it has also been observed that complete thermal-hydraulic analyses and bases have not been developed which would ensure that operators have been given the necessary information to respond to a complicated event involving steam generation in the RCS, including one following a station blackout. A number of important considerations relating to alternate decay heat removal were not found in

training literature nor plant procedures. These are discussed in Section 6.6.1.2.

Instrumentation. Most licensees have generally responded appropriately to GL 88-17 by providing two independent RCS level indications, two independent measurements of core exit temperature, the capability to continuously monitoring RHR system performance, and visible and audible alarms. However, wide variability exists among sites in the quality of installations and controls for using them, as discussed below.

- Many operators were unaware that core temperature cannot be inferred from measurements in the RHR system when the RHR pumps are not running, and sometimes core exit thermocouples have not been kept operable even though the vessel head was installed.

- Potential problems associated with water level indications have been observed, including damaged or incorrectly installed instrument tubing (or both), lack of independence, and poor maintenance.

- At some plants, the RHR system is not being monitored for problems that foreshadow system failure.

6.6.1.2 Alternate Residual Heat Removal Methods

In response to the incident investigation team's report of the loss of ac power at the Vogtle plant (NUREG-1410), the staff, with the assistance of the Idaho National Engineering Laboratory, has conducted in-depth studies of passive, alternate methods of RHR heat removal that could potentially be used when the RHR system is unavailable. The initial study (EGG-EAST-9337) identified fundamental passive cooling mechanisms that could be viable for responding to an extended loss of RHR and evaluated plant conditions and procedural actions that could be used to exploit those mechanisms, as well as problems in such exploitation. The important cooling processes include gravity drain of water from the RWST into the RCS, core water boiloff, and reflux cooling. A second study (published in April 1992 NUREG/CR-5820) examined the transient response of a PWR with U-tube steam generators following a loss-of-RHR event using the RELAP5/MOD3 reactor analysis code with a model modified for reduced inventory conditions. The significant findings from these studies are discussed below.

Gravity Drain From the Refueling Water Storage Tank. Most, but not all, PWRs are theoretically capable of establishing a drain path between the RWST and the RCS. However, the relative elevation difference between the RWST and the RCS, which determines how much water is available, can vary significantly from plant to plant. Under ideal conditions for a spectrum of plants studied, RWST feed-and-bleed of the RCS could maintain flow to the vessel and remove decay heat for as little as 0.4 hour for one plant to as much as 18 hours for another, assuming the loss of RHR occurred 2 days after shutdown; for unthrottled flow, the times are 0.2 hour and 5.2 hours.

Gravity Feed From Accumulators or Core Flood Tanks. The limited liquid contents in accumulators or core flood tanks makes their use of marginal value in terms of long-term core cooling. However, if properly controlled, water flow from accumulators can provide core cooling for several hours following an event occurring 2 days after shutdown. From the perspective of operators trying to restore normal cooling system or source of ac power, this amount of time is significant .

*Reflux Cooling .*Initiation of reflux condensation cooling depends on the ability of steam produced by core boiling to reach condensing surfaces in the steam generator U-tubes. During a plant shutdown condition, the reactor coolant level may be at reduced inventory with air or nitrogen occupying the upper volumes of the primary system. This air inhibits steam flow from the reactor vessel to the steam generator U-tubes. Important aspects of reflux initiation are (1) the initial reactor coolant water level, (2) the need to establish and preserve horizontal stratification of the liquid in the hot legs, (3) the primary system pressure needed to establish a sufficient condensing surface, and (4) the possible need for draining or venting the primary system in order to obtain a stable reflux cooling mode at an acceptable pressure.

The ability to remove decay heat through one steam generator by reflux condensation following a loss-of-RHR event during reduced inventory operation represents an alternative way to remove decay heat, one that does not require adding water to keep the core covered with a two-phase mixture. In many instances, nozzle dams are installed in the hot-leg and cold-leg penetrations to one or more steam generators, and the reactor vessel head is installed with air in the unfilled portion of the RCS above the water level. Should the RHR system fail, the peak pressure and temperature reached in the RCS are important since the nozzle dams must be able to withstand these conditions to prevent a loss-of-coolant accident. Failure of a hot-leg nozzle dam would create a direct path to the containment through an open steam generator manway. Such an event could also result in peak RCS pressures sufficient to cause leakage past the temporary thimble seals used to isolate the instrument tubes. These thimble seals are used during plant outages while nuclear instruments are retracted from the reactor (see NUREG-1410).

Analyses were performed in the NUREG/CR-5820 study to identify the time to core uncovery due to the failure of the hot-leg nozzle dam with the manway removed from the steam generator inlet plenum. Nozzle dams were assumed to fail at 25 psi (172 kPa). The actual failure pressure is not well known and likely varies among differ-

ent designs. An analysis was also performed to determine the time to core uncovery if water was lost via guide tubes that connect to the bottom of many reactor vessels.

The results of the analyses are as follows:

- Analyses of the loss of the RHR system from mid-loop operation at 1 day and 7 days following shutdown reveal that the RCS can reach peak pressures in the 25-psig (172–kPa) range when a single U-tube steam generator is used for RHR. Moreover, RCS peak pressure is insensitive to decay heat level or to the time of loss of RHR system following shutdown.

- Additional analyses of the use of U-tube steam generators for RHR show that RCS peak pressures approach 80 psig (552 kPa) with initial RCS water levels above the top elevation of the hot leg. At these higher water levels, calculations indicate that fluid expansion fills the steam generator tubes with sufficient liquid to prevent RHR until pressures reach 80 psi (552 kPa) or until sufficient primary to secondary temperature difference is established. Peak RCS pressure is, therefore, sensitive to the initial liquid level at the time the RHR system is lost.

- Since RCS pressures near the design conditions for nozzle dams and temporary thimble seals can be attained, the successful use of the steam generators as an alternative RHR mechanism is not assured. The loss of the RHR system with initial RCS water levels above the top of the hot leg suggests that using the steam generators as an alternative means of decay heat removal will result in sufficient pressure to challenge the integrity of temporary boundaries in the RCS.

- Analyses of the failure of the RCS temporary boundaries (i.e., nozzle dams and thimble seals) or openings such as the safety injection line demonstrate that if the RHR system fails within the first 7 days following shutdown, there is very little time (i.e., about 30 to 90 minutes) to prevent core uncovery under worst core condition involving a nozzle dam failure.

6.6.2 Boiling-Water Reactors

During a normal shutdown, initial cooling is accomplished by using the main turbine bypass system to direct steam to the main condenser, and by using the condensate and feedwater systems to return the coolant to the reactor vessel. The circulating water system completes the heat transfer path to the ultimate heat sink. This essentially is the same heat transport path as is used during power operation, except that the main turbine is tripped and bypassed and the steam, condensate, and feedwater systems are operating at a greatly reduced flow rate. When the steam and power conversion system is not available, high-pressure shutdown cooling is achieved by isolation condensers (early BWRs) or by the reactor core isolation cooling (RCIC) system (later model BWRs). No BWRs have both isolation condensers and an RCIC system.

The RHR system provides for post-shutdown core cooling of the RCS after an initial cooldown and depressurization to about 125 psig (862 kPa) by the steam and power conversion system, the isolation condensers, or the RCIC system. Early BWRs have dedicated RHR systems that are separate from the low-pressure ECCS subsystems. Later model BWRs have multi-mode RHR systems that perform the shutdown cooling function as well as a variety of ECCS and primary containment cooling functions. The RHR shutdown cooling suction line is opened to align the suction of the RHR pumps to a reactor recirculation loop on the suction side of an idle recirculation pump. Flow is established through the RHR heat exchangers, and the primary coolant is then returned to the reactor vessel via a recirculation line (on the discharge of an idle recirculation pump) or a main feedwater line (later model BWRs only). The RHR heat exchangers transfer heat to the RHR service water system. The RHR service water system is a single-phase, moderate-pressure system that is dedicated to providing cooling water for the RHR heat exchangers. In later model BWRs (BWR/5s and BWR/6s), RHR cooling is supplied by an essential service water system that also provides cooling for other safety-related components. In either case, the service water systems may operate on an open, closed, or combined cycle. The service water and the circulating water systems may operate on different cooling cycles (i.e., a closed-cycle service water system and an open-cycle circulating water system).

Because of the relatively high discharge pressure of the RHR service water pumps (about 300 psid (2068 kPa)), the service water system can be used in an emergency to flood the BWR core or the primary containment. This capability is implemented by opening the cross-tie between the service water system and the RHR return line to the RCS. In a multi-mode RHR system, this return line branches to the reactor vessel, the suppression pool, and the drywell.

Loss of Residual Heat Removal Capability

As indicated in Chapter 2, the frequency and significance of precursor events involving reduction in reactor vessel water level or loss of RHR (or both) in BWRs have been less than for PWRs. One reason for this is that BWRs do not enter a reduced inventory or midloop operating condition as do PWRs. Another reason is that a reduction in reactor vessel (RV) water level will normally be terminated by the primary containment and RV isolation system before the level falls below the suction of the RHR pumps.

Should RHR be lost, operators can usually significantly extend the time available for recovery of the system by adding water to the core from several sources, including condensate system, low-pressure coolant injection (LPCI) system, core spray (CS) system, and control rod drive (CRD) system. Adding inventory raises water to a level that can support natural circulation. In the event that RHR cannot be recovered in the short term, alternate RHR methods covered by procedures are normally available. The particular method selected will depend on the plant configuration and the decay heat load. If the RV head is tensioned, the reactor pressure vessel (RPV) is first allowed to pressurize and then steam is dumped to the suppression pool via a safety-relief valve (SRV), and makeup water is provided by one of the water sources listed above. If the condenser and condensate system are available, decay heat can be removed by dumping steam to the condenser and adding makeup water from the condensate and feedwater system. If the vessel head is detensioned, decay heat must be removed without the RPV pressurized. For some BWRs, this requires opening multiple SRVs to dump steam to the suppression pool and cooling the suppression pool by recirculating water using the CS or LPCI pumps. For all cooling methods involving the suppression pool, suppression pool cooling must be initiated in sufficient time to prevent suppression pool temperature from becoming so high that the pumps lose net positive suction head. If the RPV head is removed and the main steamline plugs are put in place, the preferred method of RHR is to flood the reactor cavity and place the fuel pool cooling system in operation or utilize the reactor water cleanup system. A second undesirable, but nevertheless effective, alternative is to boil off steam to the secondary containment and add makeup water from any source capable of injecting water at a rate of a few hundred gallons per minute. As discussed in Section 6.9.1, this method of RHR can lead to failure of the secondary containment.

The findings of the accident sequence precursor analysis discussed in Chapter 2 indicate that BWRs experience fewer and less severe loss-of-RHR incidents than PWRs. In addition, the review of BWR alternate RHR methods indicates significant depth and diversity. For these reasons, the staff concludes that loss of RHR in BWRs during shutdown is not a significant safety issue as long as the equipment (pumps, valves, and instrumentation) needed for these methods is operable and clear procedures exist for applying the methods.

6.7 Temporary Reactor Coolant System Boundaries

In the course of the evaluation, the staff identified and examined plant configurations used during shutdown operations involving temporary seals in the reactor coolant system. This includes freeze seals that are used in a variety of ways to isolate fluid systems temporarily, temporary plugs for nuclear instrument housings, and nozzle dams in PWRs. The staff has noted instances in which failure of these seals, either because of poor installation or an overpressure condition, can lead to a rapid non-isolable loss of reactor coolant. This concern is of special importance in PWRs because the emergency core cooling system (ECCS) is not designed to automatically mitigate accidents initiated at pressures below a few hundred psig and is not normally fully available for manual use during these conditions. In BWRs, the ECCS is required to be operable during cold shutdown, and during refueling when there is fuel in the reactor vessel and the vessel water level is less than 23 feet above the reactor pressure vessel flange. In addition, the ECCS is actuated automatically when water level is low in the reactor vessel.

6.7.1 Freeze Seals

Freeze seals are used for repairing and replacing such components as valves, pipe fittings, pipe stops, and pipe connections when it is impossible to isolate the area of repair any other way. Freeze seals have been used successfully in pipes as large as 28 inches (71 cm) in diameter. However, as a result of inadequate use and control, some freeze seals have failed in nuclear power plants, and some of the failures have resulted in significant events. This has raised a question regarding the adequacy of 10 CFR 50.59 safety evaluations of freeze seal applications.

To assess problems associated with freeze seals, the staff reviewed the operational experience on freeze seal failures, safety-significant findings on freeze seal failures, industry reports on freeze seal use and installation, and the applicability of industry guidance (NSAC-125) for performing safety evaluations on freeze seal applications.

6.7.1.1 Operational Experience on Freeze Seal Failures

River Bend, 1989. Failure occurred in a freeze plug (used in a 6-inch [15 cm] service water line to allow inspection and repair work on manual isolation valves to a safety-related auxiliary building cooler). The failure caused a spill of approximately 15,000 gallons (56,781 L) of service water into the auxiliary building and caused the loss of non-safety-related electrical cabinets (i.e., shorting and an electrical fireball damaged cabinets and components). Draining water also tripped open a 13.8-kV supply breaker, leading to loss of the RHR system, spent fuel pool cooling system, and normal lighting in the auxiliary and reactor buildings. The leak was isolated in 15 minutes and the RHR system restarted in 17 minutes.

Oconee 1, 1987. Approximately 30,000 gallons (113,562 L) of slightly radioactive water leaked into various areas of the auxiliary building, and a portion drained beyond the site boundary when a freeze plug (used to facilitate re-

placement of a 3-inch-diameter [7.6 cm] section of low-pressure injection piping) failed.

Brunswick 1, 1986. Failure of a freeze seal (used in the discharge piping of the control rod drive system pump 1A) caused hydraulic perturbation to a high-level/turbine trip instrument, resulting in a feed pump trip and subsequent automatic scram at 100-percent power.

The freeze seal failure at River Bend prompted a visit by an NRC augmented inspection team (AIT) to perform an onsite inspection shortly after the event. The AIT found

* inadequate control of freeze seal work

* lack of training for personnel performing the work

* lack of awareness by plant personnel of the potential for freeze seal failure

* flooding that exceeded the design capacity of the floor drain system

* no damage to safety-related equipment

A 10 CFR 50.59 safety evaluation of the freeze seal operation was not performed. The plant operating procedure was subsequently revised to include corrective measures for freeze seal installation and control. However, the licensee included no statement to ensure or require that a 10 CFR 50.59 safety evaluation be performed before allowing use of a freeze seal.

In regard to the incident that occurred at Oconee, the NRC cited the utility for inadequate freeze seal procedures. A review of the licensee's freeze seal "safety evaluation checklist" found that the checklist questions were similar to 10 CFR 50.59 questions. However, the checklist was not processed through the licensee's safety committee, as would have been done for a formal 10 CFR 50.59 safety evaluation.

Information Notice 91–41, "Potential Problems With the Use of Freeze Seals," identified potential problems related to the freeze seal in PWRs and BWRs, specifically including both the River Bend and Oconee 1 incidents. The information notice indicated that freeze seal failure in a PWR reactor boundary system could result in immediate loss of primary coolant. In BWRs, failure of a freeze seal in a system connected to the vessel's lower plenum region, such as the reactor water cleanup (RWCU) system, could result in the water level in the reactor vessel falling below the top of the active fuel. The estimated time for this to occur is less than 1 hour if the seal failed completely and makeup water was not added to the reactor. The information notice indicated concerns that freeze seal failures in secondary systems can also be significant because of the potential for consequential failures, such as the loss of RHR in the River Bend event. The information notice identified procedural inadequacies that resulted in a failure to install and monitor a temperature detection device, and a lack of personnel training in the use of freeze seals. Other important considerations identified in the notice included: "examining training, procedures, and contingency plans associated with the use of freeze seals, and evaluating the need for and availability of additional water makeup systems and their associated support systems." No specific statement was included regarding the applicability of a 10 CFR 50.59 safety evaluation.

6.7.1.2 Industry Reports on Use and Installation of Freeze Seals

In February 1989, the Electric Power Research Institute issued EPRI NP-6384-D, "Freeze Sealing (Plugging) of Piping," to guide nuclear power plant maintenance personnel in evaluating the use of freeze seals. The guide cautioned personnel on the use of freeze seals and discussed contingency plans should freeze seals fail.

The Battelle Columbus Laboratories issued a final report, "Development of Guidelines for Use of Ice Plugs and Hydrostatic Testing," in November 1982; the report discussed the potential hazards associated with ice plugs and gave guidelines for plug slippage, restraint, pressure, impact loads, and stress arising from handling. Defects and personnel safety were also discussed.

6.7.1.3 NSAC-125, "Industry Guidelines for 10 CFR 50.59 Safety Evaluations"

NSAC-125, issued in June 1989 by the Nuclear Management and Resources Council (NUMARC), gave the industry guidelines for performing 10 CFR 50.59 safety evaluations. The document provided industry guidance on the thresholds for unreviewed safety questions, the applicability of 10 CFR 50.59, and the procedures for performing 10 CFR 50.59 safety reviews for facility changes, tests, or experiments at nuclear power stations. The staff's review of NSAC-125 identified the following as appropriate guidance for the applicability of the 10 CFR 50.59 safety evaluation to the use of freeze seals as temporary modifications and the application of the 10 CFR 50.59 determination of whether an unreviewed safety question exists for the freeze seal installation: "Temporary changes to the facility should be evaluated to determine if an unreviewed safety question exists. Examples of temporary modifications include jumpers and lifted leads, temporary lead shielding on pipes and equipment, temporary blocks and bypasses, temporary supports, and equipment used on a temporary basis."

Although the use of freeze seals as a temporary block is not specifically identified, freeze seals perform the "temporary block" function and, therefore, the staff finds they

conform with the NSAC-125 definition of "temporary modifications."

6.7.1.4 Results and Findings

- For BWRs, failure of a freeze seal in a system connected to the vessel's lower plenum region such as the RWCU system, could cause the core to become uncovered in less than 1 hour if the seal failed completely and the ECCS failed to perform its intended function of adding makeup water to the reactor.

- NSAC-125, industry guidance for applying 10 CFR 50.59, covers temporary modifications, but does not discuss freeze seals specifically.

- Temporary modifications using freeze seals are not being evaluated per 10 CFR 50.59.

- Industry guidance exists for using freeze seals with contingency plans.

- Operating experience indicates that freeze seal failures could constitute safety problems.

6.7.2 Thimble Tube Seals

The arrangement of the incore instrumentation assemblies in many PWRs may be visualized by considering one end of an approximately 1-inch (2.5-cm)-diameter tube as welded to the bottom of the reactor vessel and the other end welded to the seal table. This tube provides a penetration into the reactor from below, with the opposite end containing a high-pressure seal during power operation. This "guide" tube is a permanent part of the reactor coolant system pressure boundary.

A thimble tube that has a closed end is inserted into the guide tube, closed end first, and is pushed through the guide tube until it extends up into the reactor core. The thimble tube is then sealed to the guide tube by a high-pressure, Swagelok-type fitting at the seal table, thus forming a watertight assembly with the area between the tubes containing reactor coolant system water and the inside of the thimble tube open to the containment building. The space between the tubes is subjected to reactor coolant system pressure during power operation.

Preparation for refueling involves withdrawing the thimble tubes out of the core. Thus, the normal seal between the Swagelok-type thimble tube and the guide tube at the seal table must be opened.

Once the thimble tube is withdrawn from the core region, the annular gap is closed, often by a temporary seal comprising split components and rubber gaskets. Temporary thimble tube seals have a typical design pressure of 25 psi (172 kPa), so that a significant overpressurization could cause them to fail. This would cause a leak that is effectively in the bottom of the reactor vessel.

The thimble tubes in plants designed by Babcock and Wilcox (B&W) terminate in an "incore instrumentation tank" that is open at the top, at the refueling floor level, with the bottom at roughly reactor vessel flange level. No temporary seals are used and the tank fills with water (or is filled) so that tank and refueling cavity water level remain the same. There can be times during typical refueling outages when the tank is open to the containment at the bottom and when some of the guide tubes are empty, thus providing a potentially significant flow path between the bottom of the reactor vessel and the incore instrumentation tank as well as to the containment.

Most units designed by Combustion Engineering (CE) do not use such bottom-entering incore instrumentation as described above. The staff understands that the few that do, use a B&W-type arrangement to terminate the tubes in the refueling cavity rather than a separate tank.

Analysis of Leakage Via Instrument Tube Thimble Seal Failure

Leakage due to instrument tube thimble seal failure in a Westinghouse-designed plant was analyzed to determine how long it takes to uncover the core when one steam generator is used to remove decay heat following a loss of RHR. This analysis is part of the transient thermal-hydraulic analysis of the loss of RHR in a PWR discussed in Section 6.6.1.2.

Thimble seal failure in the instrument tubes was assumed to occur when system pressure reached 20 psig (138 kPa). This value was chosen to investigate the consequences of failure of the thimble seals and may not reflect actual failure pressures for seals. For this analysis, it was assumed that there were 58 thimble seals and all of these seals fail, once the assumed failure pressure is achieved. The break flow area selected for the analysis was based on the cross-sectional area of the thimble tube. This bounds the actual area which is more accurately represented by the annular area between the thimble tube and guide tube. The failure was assumed to be located at the seal table, which is at the elevation of the reactor vessel flange for the plant modeled. The tubes are connected to the vessel at the bottom of the lower head and are collected at the seal table resulting in an elevation difference between these two locations of about 22.5 feet (6.7 m).

The RCS was initialized with water at 90 °F (32 °C) at a level at the centerline of the hot and cold legs. One steam generator was available. Air at 90 °F (32 °C) and 100-percent relative humidity is present in all volumes above the centerline of the hot and cold legs. The decay heat power

level corresponding to 1 day after shutdown was conservatively assumed for the three-loop plant modeled in this analysis 10,900 Btu/s (11.5 MW).

Thimble seal failure is predicted to occur at about 1.6 hours after the RHR system is lost. Core uncovery in this conservative analysis is predicted to occur about 20 minutes later if makeup is not provided.

6.7.3 Intersystem Loss-of-Coolant Accidents in PWRs

Intersystem loss-of-coolant accidents (ISLOCAs) are a class of accidents in which a break occurs in a system connected to the reactor coolant system (RCS), causing a loss of RCS inventory. This type of accident can occur when a low-pressure system is inadvertently exposed to high RCS pressures beyond its capacity. During shutdown operations, this would most likely involve the RHR system that interfaces directly with the RCS via the hot leg. Because of a higher primary pressure present in PWRs, as compared to BWRs, and the more significant precursor events in PWRs, there is greater concern for ISLOCAs in PWRs. However, in all cases, the ISLOCAs of most concern are those that can discharge RCS fluid outside the reactor containment building. In those ISLOCAs, the lost RCS inventory cannot be retrieved for long-term core cooling during the recirculation phase.

The principal cause for an ISLOCA in a PWR during shutdown is overpressurization of the RHR system. Inspections and analyses conducted by the staff indicate that in PWRs this could be caused by human errors, notably during testing and maintenance, or by an extended loss of decay heat removal capability combined with a failure of isolation valves between the RCS and RHR system to close, such as during a station blackout.

The consequences of an ISLOCA during shutdown are not expected to be significantly different from those of other shutdown-related loss-of-RHR accidents and loss-of-coolant accidents discussed previously in this chapter. This is because these accidents may very well involve an open containment, and also lack of recirculation capability due to failure of low-pressure injection pumps or a blocked containment sump.

In light of this, the staff has concluded that the risk from an ISLOCA during shutdown can be reduced significantly by (1) improving training in pertinent operations and procedures, (2) establishing contingency plans that provide for conservation and replenishment of RCS inventory in the event of an accident, and (3) planning and conducting shutdown operations in a way that maximizes availability of electric power sources.

6.8 Rapid Boron Dilution

The staff, with the assistance of Brookhaven National Laboratory (BNL), has completed a study of rapid boron dilution sequences which might be possible under shutdown conditions in PWRs; the NRC issued this report as NUREG/CR-5819. Concerns relating to rapid boron dilution during a PWR startup were raised by the French regulatory authority in its shutdown PRA study. These sequences are the result of a two-step process. In the first step it is assumed that unborated (or highly diluted) water enters the normally borated reactor coolant system (RCS) while the reactor coolant is stagnant in some part of the primary system. This diluted water is then assumed to accumulate in this region without significant mixing. The second step is the startup of a reactor coolant pump (RCP) so that the slug of diluted water will rapidly pass through the core with the potential to cause a power excursion sufficiently large to damage the core. Other variations to this two-step process include (1) having the slug forced through the core by the inadvertent blowdown of an accumulator and (2) having a loop isolated using loop stop valves and, after the loop becomes diluted, opening the loop stop valves while the RCPs are running.

6.8.1 Accident Sequence Analysis

This study considered both probabilistic and deterministic aspects of the problem and focused on what is expected to be the most likely of the several sequences that were identified as leading to a rapid dilution. This particular sequence starts (see NRC Information Notice 91-54) with the highly borated reactor being deborated as part of the startup procedure. The reactor is at hot conditions with the RCPs running and the shutdown banks removed. Unborated or diluted water is being pumped by charging pumps from the volume control tank into the cold leg. The initiating event is a loss of offsite power (LOOP). This causes the RCPs and the charging pumps to trip and the shutdown rods to scram. The charging pump comes back on line quickly when diesel generators start up. Charging continues until the volume control tank is empty and it is assumed that there is little mixing with the water in the RCS so that a region of diluted water accumulates in the lower plenum. It is then assumed that power is recovered so that the RCPs can be restarted. This is assumed to occur after sufficient diluted water has accumulated so that the slug of diluted water which then passes through the core has the potential to damage the fuel.

The probabilistic analysis was done for this scenario for a CE plant (Calvert Cliffs), a B&W plant (Oconee), and a Westinghouse (W) plant (Surry). The reactor systems and operating procedures involved in the scenario were reviewed, and accident event trees were developed. The analysis focused on the specific arrangement of the makeup and letdown systems and the chemical and volume control system. The startup and dilution procedures were important, as were the procedures to recover from a LOOP.

The initiating frequency of the scenario was considered for both refueling and non-refueling outages and varied from 2.0×10^{-4} per reactor-year to 6.0×10^{-8} per reactor-year, depending on the reactor. The probability that the injected water would cause a region of diluted water before an RCP was started was treated as a time-dependent function. It was assumed that there was no mixing of injectant after refueling and sufficient mixing after a non-refueling outage to reduce the probability of core damage by a factor of 0.5. However, the core-damage probability is not constant in time because it takes time to accumulate sufficient diluted water, and because, after emptying the volume control tank, the suction from the charging pump switches to a source of highly borated water. The time dependence of the probability of restarting an RCP was also taken into account. The resulting core-damage frequency was found to vary from 1.0×10^{-5} to 3.0×10^{-5} per reactor-year.

6.8.2 Thermal-Hydraulic Analysis for the Event Sequence

A key assumption in the probabilistic analysis is that the injectant does not mix with the existing water in the RCS so that a diluted region accumulates in the lower plenum. This assumption was tested by using mixing models to determine to what extent charging flow mixes with the existing water when it is injected into a loop that is either stagnant or at some low natural circulation flow rate insufficient to provide complete mixing. These mixing models are based on the regional mixing models that were developed to understand the thermal mixing of cold injectant into the "cold" leg which is at a much higher temperature. The thermal mixing problem was originally of interest for the problem of pressurized thermal shock.

The regional mixing model has been utilized to calculate the boron concentration in the mixed fluid when the unborated, cold, injected water mixes with the hot water in the cold leg which is taken to have a boron concentration of 1500 ppm. The model specifically considers the mixing region near the point of injection and at the end of the cold leg where the flow is into the downcomer, and ignores mixing in the downcomer or lower plenum.

The model was applied to the Surry and Calvert Cliffs plants under the assumption of no loop flow. The finding was that there is considerable mixing so that the water in the lower plenum would have a boron concentration that is only 400–600 ppm less than that originally in the core. On the basis of the neutronic calculations explained below, this is insufficient to cause a power excursion when an RCP is restarted, unless the core design results in both a very low Doppler reactivity coefficient and a very low shutdown bank reactivity worth. It is difficult to generalize these results as they are dependent on specific plant parameters defining the loop geometry and the charging flow.

6.8.3 Neutronics Analysis

The neutronics of this problem was studied to understand the consequences of having a slug of diluted water pass through the core. In order to do simple scoping calculations, the staff took a synthesis approach. This approach combines steady-state, three-dimensional core calculations of boron reactivity worth under different configurations with point kinetics calculations of the resulting power transient.

The steady-state calculations were done with the NODEP-2 nodal code. The output from these calculations is the static reactivity worth of a diluted slug as a function of position of the slug as it moves through the core. The two basic shapes that have been considered are a semi-infinite slug (step function) and a finite slug (rectangular wave function) with a volume of 535 cubic feet (15 m³). The calculations were done with different dilutions, relative to the 1500 ppm assumed as the initial state of the core. In addition to a radially uniform slug, two other geometries were considered. In one, the slug was localized in the center 49 assemblies, and in another, the slug was found at two peripheral locations affecting 50 assemblies. The calculations provided not only reactivity versus position of the leading edge of the slug but also Doppler weight factors for use in the kinetics calculations.

The dynamics calculations included the neutron kinetics as well as a simple fuel rod conduction model to calculate a more accurate fuel temperature than would be obtained by making an adiabatic assumption. The calculated peak fuel enthalpy was used as the criterion to judge whether fuel had been damaged. If the calculated peak fuel enthalpy exceeded 280 calories per gram (1172 J/gram), catastrophic fuel damage involving a change in geometry was assumed to occur. The peak fuel enthalpy was calculated using the time-dependent power and a power peaking factor taken from the static three-dimensional calculation at the condition corresponding to the time of the peak power.

The results show that fuel damage could occur if the boron concentration in a semi-infinite slug is reduced to 750 ppm, corresponding to an equal mixing of injected water at 0 ppm and reactor coolant at 1500 ppm. These results are dependent on the worth of the shutdown banks and on the Doppler reactivity coefficient; calculations were done to determine this sensitivity.

6.8.4 Other Analyses

Transient calculations somewhat similar to these studies have been done by several other groups. Two examples follow:

(1) Westinghouse (S. Salah et al.) performed calculations for a situation wherein the loop stop valves are both cold (down to 70 °F [21 °C] from 547 °F [286 °C]) and completely unborated due to an unknown mechanism. W used a three-dimensional neutron kinetics analysis to assess the core response when the loop stop valves were assumed to open while the RCPs were running. All rods were assumed to be initially out of the core and, hence, the worth of the scram reactivity (not including the assumed "stuck rod") would be about 6- or 7-percent delta-k. The result, for an initial 1500-ppm boron concentration, was (a) integrated core power not above normal core average power, but (b) localized fuel damage in the cold, unborated, stuck rod core region, involving only about 3 percent of the fuel and "not sufficient energy release to break the integrity of the primary system."

(2) Calculations performed as part of a thesis (S. Jacobson) examined similar transients with various dilution scenarios. The steam generator tube rupture/accumulation of a diluted region during primary pump shutdown/rapid core dilution following pump turn-on was the most significant event found in the study. The conclusion drawn from this study was that the fuel failure criterion (similar to that used in the BNL studies above) is not exceeded.

The review and analysis of rapid boron dilution events during shutdown appear to indicate that core damage may occur for assumed extreme sets of event parameters, including a necessary assumption of minimal mixing of diluted and borated water, and may occur with a frequency of the order of 10^{-5} per reactor-year. These events can be prevented by the use of appropriate procedures which anticipate the possibility of dilution in various recognized situations and prevent it, or prevent the inappropriate starting of pumps until suitable mixing procedures are carried out.

6.9 Containment Capability

6.9.1 Need for Containment Integrity During Shutdown

The NRC staff performed scoping calculations of core heatup for a Westinghouse four-loop PWR to allow assessment of containment response and a potential release. For loss of RHR during midloop operations, the time to heat the core to boiling was calculated as 8 minutes. Once boiling began, the reactor vessel level could decrease to the top of the active fuel in as little as 50 minutes. This calculation assumed that the reactor had operated for a full cycle and had been shut down for 48 hours. Additionally, 35 percent of the reactor coolant inventory between the top of the active fuel and the middle of the hot leg was assumed to spill from the RCS.

PWRs have containment structures that are classified as large dry, subatmospheric, or ice condenser. For any of these containment designs, the reestablishment of containment integrity before core damage occurs is important for reducing offsite doses. The effect of a primary containment in reducing the offsite dose consequences is evaluated by comparing what might occur if the containment were open to what might occur if the fission products remained within the closed containment. An open containment would allow direct release of steam and fission products to the atmosphere; holdup in the containment would allow plateout and decay to occur.

Offsite dose consequences from a postulated severe accident were evaluated with and without a containment in the NRC "RTM 91: Response Technical Manual," NUREG/BR-0150. RTM-91 evaluated offsite dose at a distance of 1 mile from a typical site for varying degrees of core heatup and damage. The values used there were based on the assumption that the release occurs immediately after shutdown. In one case, the dose was evaluated for an accident causing damage only to the fuel cladding with release of the volatile fission products stored in the fuel pin gap space. The dose rate from further heating included the release of the volatile fission products retained in the grain boundary regions within the fuel pellets and, finally, release following a postulated core melt was considered. Without the benefit of containment retention, the doses 1 mile (1.6 km) from the plant would be high, ranging from 20 rem (0.2 Sv) (whole body) and 2000 rem (20 Sv) (thyroid) for a gap release to 1000 rem (10 Sv) (whole body) and 100,000 rem (1000 Sv) (thyroid) for a postulated core melt.

A release 48 hours after shutdown would also have severe consequences since most of the dose to the thyroid would come from inhaling iodine-131. Iodine-131 has a half-life of 8.1 days for a dose reduction by a factor of 0.84 after 48 hours. The whole-body dose would be somewhat more affected by a prior shutdown of 48 hours since short-lived isotopes make up about 80 percent of the whole-body dose following an immediate release. The whole-body dose 1 mile (1.6 km) from the plant would be about 200 rem (2 Sv) considering 48-hour decay. This would come principally from iodine-131 with its 8.1-day half-life. Further retention of the fission products prior to release would cause the offsite dose to be reduced by about 97 percent of the initial release value, with long-lived cesium isotopes as the principal contributors to dose. These estimates assumed release of 25 percent of core iodine and 1 percent

of particulates. The evaluations are appropriate for large dry PWR containments, subatmospheric containments, and ice condenser containments for which the ice bed was bypassed by the escaping steam. For releases through the ice bed, reduction factors of between 0.3 and 0.5 are expected.

The effect of holdup and plateout in the containment on offsite dose was determined in RTM-91 to be significant. With a 24-hour holdup in the containment and with design leakage assumed, calculated offsite doses are reduced to 5×10^{-5} rem (5×10^{-3} Sv) (whole body) and 4×10^{-3} rem (4×10^{-1} Sv) (thyroid) for the gap release case and 0.002 rem (0.02 mSv) (whole body) and 0.2 rem (2 mSv) (thyroid) for the core-melt case. Thyroid and whole-body doses are further reduced by factors of 5 and 3, respectively, if the containment spray was operated during the event. Doses would, of course, be increased by any subsequent containment failure and revaporization of fission products that might occur following a hypothetical accident involving severe core damage.

BWRs are not typically operated in a reduced inventory condition as are PWRs. However, 2 days into an outage, a BWR/4 (such as Browns Ferry) may have as little as 205 inches (521 cm) of reactor coolant above the top of the active fuel. If shutdown cooling were lost, boiling would begin in 28 minutes. The reactor vessel water level would be at the top of the active fuel 308 minutes later. This corresponds to a steam flow rate of 24,800 cubic feet per minute (702 m³/min) into the Mark I secondary containment with the drywell head removed for refueling.

This flow into the secondary containment could increase the internal pressure to 0.5 psig (3.4 kPa) in 5 minutes. Such pressure is significant because the secondary panels are designed to blow out at 0.5 psig (3.4 kPa), releasing steam and fission products directly to the atmosphere. The calculation to determine the time to secondary containment failure was based on an energy balance after depositing 285,000 pounds (129,273 kg) of steam into the secondary containment. The heat sink inside the secondary containment is made up of structural steel and air. No secondary system leakage was assumed.

Two other calculations were performed to determine the secondary containment's sensitivity to changes in the mass of structural steel and air inside the secondary containment. The first calculation increased the mass of steel inside the secondary containment by five times that used in the previous calculation. This increased the amount of time for the secondary containment to reach 0.5 psig (3.4 kPa) from 5 minutes to 6 minutes. The second calculation decreased the volume of the secondary containment from 4 million cubic feet (113,267 m³) to 2 million cubic feet (56,633 m³). That resulted in decreasing the amount of time from 5 minutes to 3 minutes for the secondary containment to reach 0.5 psig (3.4 kPa). This sensitivity study was necessary because secondary containment designs and sizes vary from plant to plant.

RTM–91 also evaluated offsite doses at a distance of 1 mile from a typical BWR site for varying degrees of core heatup and damage. If the drywell head were removed, the release could go directly into the secondary containment and through the blowout panels for Mark I and II secondary containments, bypassing standby gas treatment. As in the PWR evaluation, the dose was calculated for releases from three cases: the fuel pin gap space, the grain boundary, and core melt. The BWR doses would range from 20 rem (whole body) (0.2 Sv) and 2000 rem (20 Sv) (thyroid) for a gap release to 1000 rem (10 Sv) (whole body) and 100,000 rem (1000 Sv) (thyroid) for a postulated core melt. These are the same doses listed for the PWR case.

RTM–91 Table C–3 gives a reduction factor of 0.01 for dry low-pressure flow and 1.0 for wet high-pressure flow through the standby gas treatment system filters. Considering the fact that 24,800 cubic feet per minute (702 m³/min) of saturated steam is being deposited inside the secondary containment and a typical standby gas treatment exhaust fan is only rated for 5000 cubic feet per minute (142 m³/min), the flow through the standby gas treatment system will be closer to the wet high-pressure case and the dose will not be significantly reduced.

6.9.2 Current Licensee Practice

GL 88–17 was issued to PWR licensees and required, among other things, implementation of procedures and administrative controls that reasonably assure that containment closure will be achieved before the time that RPV water level would drop below the top of the active fuel following a loss of shutdown cooling under reduced inventory conditions. The NRC staff assessed whether the requirements of GL 88–17 were in place by implementing special inspections at each site under the inspection guidance in Temporary Instructions TI–2515/101 and 2515/103. The Vogtle Incident Inspection Team recognized the need to develop broader recommendations for low-power and shutdown operation. This led to the NRC staff's program to visit selected plant sites undergoing low-power/shutdown operation (see Chapter 3). The staff also observed a variety of practices at the sites. For PWRs, the staff noted that licensees did not meet the recommendations of GL 88–17. Some licensees went beyond the recommendations of GL 88-17 by providing procedures for rapid containment closure for plant conditions other than reduced inventory.

Closure of the equipment hatch would be required for maintaining containment integrity. In one case, a polar crane would have to be used. Some licensees utilized the equipment hatch as a passageway for electrical cables and hoses. At these sites, rapid removal of this equipment was provided for by the use of quick disconnects. Some plants also provided bolt cutters and axes for contingency use. One of the sites visited demonstrated an equipment hatch closure capability requirement of within approximately 15 minutes of loss of RHR. The onsite review report noted that this was more often the exception than the rule.

Several factors are key to ensuring that the equipment hatch is closed in a timely matter. These include accounting for radiological and environmental conditions that could result from reactor coolant being boiled into the containment, addressing the number and location of closure bolts, providing for the loss of ac power, keeping tools needed for closing the equipment hatch near at hand, and finally, training and rehearsing personnel in the closure procedure. The closure of the equipment hatch in sufficient time is essential to keeping possible releases within established guidelines. These observations also apply to licensees with BWR Mark III primary containments. GL 88–17 was not sent to BWR licensees, and the onsite review report noted that these licensees have not made provisions for rapid equipment hatch closure.

A licensee, reporting a quarter-inch gap at the top of the equipment hatch when four bolts were used, found it necessary to use two more bolts to close the gap. GL 88–17 specified a no-gap criterion for hatch closure, but not every licensee confirmed that this condition was achieved. Tests or observations must be performed on internal equipment hatches to determine the location and minimum number of bolts needed to obtain an adequate closure. For external hatches, containment pressure effects on hatch closure must be considered along with the source term when evaluating the minimum number of bolts necessary to achieve an acceptable leak-tightness.

Procedures for controlling and closing containment penetrations varied widely. Some licensees did not initiate closure until temperatures exceeded 200 °F (93 °C). Above 200 °F (93 °C), boiling might begin quickly. The licensees, however, had not evaluated the in-containment environment and the ability of personnel to work in that environment to perform the necessary containment closure operations. Some plants require that the containment always be closed during midloop operations. One licensee interpreted this as meeting GL 88–17 recommendations and, therefore, did not develop procedures for rapid containment closure. Water-filled, U-pipe, loop-seal configurations found at several plants provided containment entry for electrical cables and tubing. The water-filled U-pipes were judged inadequate for withstanding containment pressure conditions that might exist following a loss of shutdown cooling.

6.9.3 PWR and BWR Equipment Hatch Designs

In order to gain a better understanding of primary containment capability in PWRs and BWRs during an accident that occurs while a plant is shut down, the staff gathered information on the design of equipment hatches from resident inspectors at U.S. plants.

The hatch survey was conducted using a questionnaire on specific equipment hatch characteristics. Answers to the questionnaire were tabulated and grouped under BWR or PWR. For BWRs, the survey asked for information on the equipment hatch that would be used only for removing a recirculation pump motor; the survey did not address removing and replacing a drywell head. The results of the survey are tabulated in Appendix B.

The majority of equipment hatches for both BWRs and PWRs were pressure-seating hatch designs (67% for BWRs, 86% for PWRs). For BWRs, the resident inspectors who were polled indicated that the equipment hatch (either recirculation pump motor or CRD hatch) would generally be removed along with the drywell head, but that removal of the equipment hatch alone was unlikely.

PWR equipment hatches consisted of 9 of the pressure-unseating type and 33 of the pressure-seating type. Of the plants surveyed, 52 required the use of ac power or compressed air (or both) to install the hatch under normal conditions, but five resident inspectors indicated that the licensee had a procedure for closing the hatch manually. Four plants with pressure-unseating hatches can use a truck-mounted crane to install the equipment hatch during a loss of normal ac power. Five PWR plants did not require the equipment hatch to be in place during fuel movement. They are Braidwood, Byron, Palisades, San Onofre 1, and Zion. These have their hatches located so that they open to the fuel handling building which has a heating, ventilation, and air conditioning system to process contaminated air during a fuel drop event.

Three PWR resident inspectors and the licensees for Catawba, McGuire, and Salem have noticed that the minimum number of bolts as specified in the technical specification is not sufficient to bring all hatch sealing surfaces into contact. A noticeable gap was present with use of the minimum number of bolts. Two licensees (Palo Verde and Summer) ran successful leak tests, an Appendix J (10 CFR Part 50) type A and a type B, with the minimum number of bolts installed. Discussion with two hatch vendors indicated that hatches have been designed so that the sealing surfaces should mate when the minimum number of bolts was installed. Ginna and Indian Point 2 have fabricated temporary closure plates that are used when the equipment hatch is removed, but tempo-

rary services are run into the containment. The Indian Point 2 temporary closure plate is rated for 3 psid (21 kPa) and has penetrations for fluid and electrical services.

6.9.4 Containment Environment Considerations for Personnel Access

6.9.4.1 Temperature Considerations

The NRC staff estimated that approximately 50,000 pounds (22,680 kg) of steam could be deposited inside the containment 1 hour after RHR in a $\underline{W}$ four-loop PWR occurring 2 days after shutdown. The steam is a result of boiling in the reactor coolant from the middle of the hot leg to the top of the active fuel, and it is assumed that 35 percent of the reactor coolant is spilled from the RCS. The staff assumed that the containment volume was 2 million cubic feet (56,633 m³) of dry air at 70 °F (21 °C) and that the containment environment after the event would consist of air and structural steel at an elevated temperature, steam, and condensed steam in the form of water. The calculation did not consider the containment fan coolers and assumed no leakage from the containment. Under these conditions, the staff expects the containment atmosphere to go from 70 °F (21 °C) and atmospheric pressure to 150 °F (66 °C) and 5.9 psig (40.7 kPa) in about 1 hour (see Figure 6.1).

This condition would be of concern because at about 160 °F (71 °C) the air is hot enough to burn the lungs. Therefore personnel inside the containment would have to be equipped with self-contained breathing apparatuses.

6.9.4.2 Radiological Considerations

Boiling of coolant within an opened reactor system following a postulated loss of shutdown cooling would release dissolved fission products within the containment atmosphere. If significant radioactivity were contained in the coolant, high-radiation-area alarms would be actuated. These are typically set at twice the background level. Health physics personnel would be expected to evacuate the containment until people could safely enter, observing the appropriate precautions and protective measures, to perform any operation required to close the containment.

To assess the radiological conditions that workers might experience while closing the containment, the NRC staff performed scoping calculations. The staff assumed that the coolant contained the expected activity for a typical operating PWR and then for a BWR as given in RTM-91. Radioactive decay was assumed to progress for 48 hours before boiling began. Iodine decay into xenon was included. The resultant concentration for PWRs was about 1/20 of the 1.0 microcurie-per-milliliter maximum

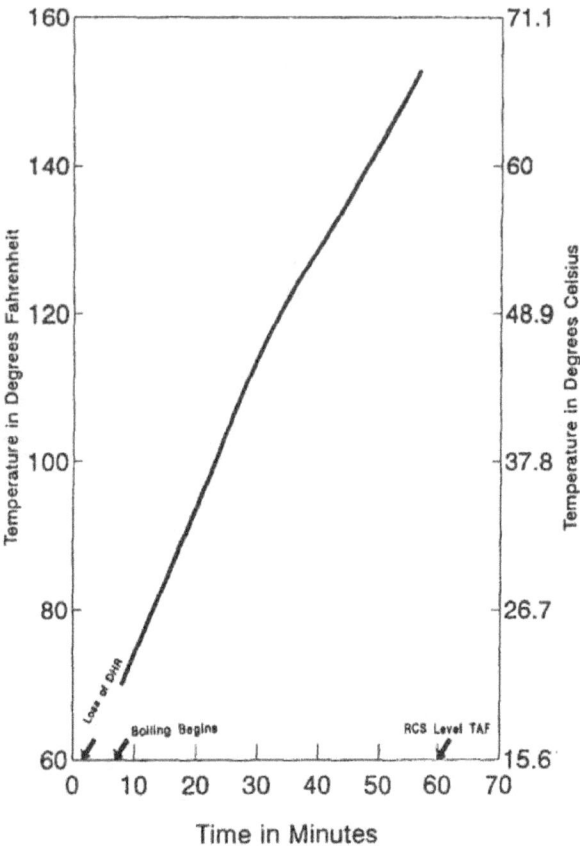

Figure 6.1 PWR Containment Temperature vs. Time

Access Training Manual which states that the risk of one Part 20 maximum permissible concentration (MPC)-hr is approximately equal to 2.5 mrem (0.025 mSv) of whole-body dose.

The resulting PWR equivalent doses are depicted in Figures 6.2 and 6.3. (These ordinarily are *conservative* because they do not include the factor-of-100 reduction discussed in the preceding paragraph.) Inhaled iodine dose in the non-respirator case was computed using soluble MPCs, whereas the respirator case was computed using the insoluble MPCs for iodine. The calculated equivalent dose increases with time and approaches asymptotic values for a pure steam atmosphere. These calculations indicate that self-contained breathing apparatus would be required for an extended stay within the containment because of the dose and humidity, since the filtration type would not function adequately in high humidity above about 106 °F (41 °C). It may be difficult to perform containment closure operations in self-contained breathing apparatus because the air supply will limit how long personnel can stay on the job. In evaluating recovery actions following a potential loss of shutdown cooling, licensees should avoid plant conditions in which steaming could occur before the containment was closed, unless reduced coolant activities or limited requirements for personnel entry indicated that the associated risk was acceptable.

Using the expected coolant activities in RTM-91 for BWRs, the calculated equivalent dose with and without respirator protection was much less than for PWRs. See Figures 6.4 and 6.5. This is because BWRs do not retain volatile fission products in the coolant. The loss of shutdown cooling with subsequent boiling was assumed to

equivalent of iodine-131 allowed in plant technical specifications. Although there is no specific requirement, PWR operators typically reduce coolant activity by two orders of magnitude using coolant cleanup systems before opening the reactor system. Additional reduction can be achieved, but the length of the outage might be increased. The scoping calculation should be considered conservative because it did not account for coolant cleanup.

The volatile fission products—noble gases and iodine—were assumed to be carried out with the boiled coolant. With these assumptions, the release of fission products to the containment was calculated concurrently with the steam released by decay heat boiling. The boiling rate was based on decay heat from a 3400-MWt plant shut down for 48 hours at the end of cycle. The steam was assumed to be mixed with the containment atmosphere (2 million cubic feet [56,633 m³]/PWR) and the mixture released through containment openings at a constant volumetric flow. Dose rates were derived from the guidance in the NRC Site

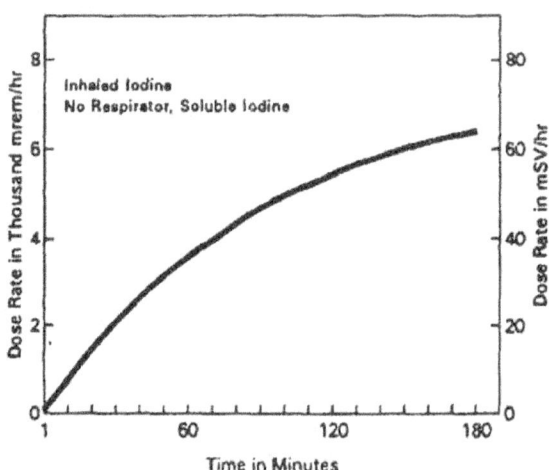

Figure 6.2 PWR Equivalent Whole-Body Dose (Inhaled Iodine)

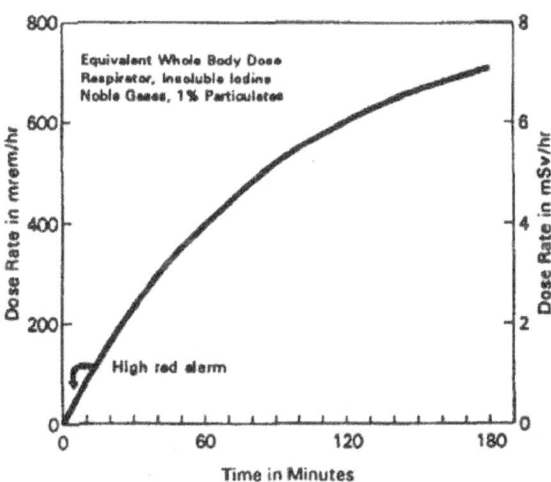

Figure 6.3 PWR Equivalent Whole-Body Dose (Noble Gases, Particulates)

occur in a typical Mark II primary containment 48 hours after shutdown with the drywell head removed. Perfect mixing was assumed in the secondary containment volume above the refueling floor (1.6 million cubic feet [45,307 m³]). Other assumptions were similar to the PWR calculation. The lower dose rates calculated for the BWR would allow for a longer stay within the secondary containment than allowed for the PWR case, and the major concern may be the steam conditions in working areas. If practical, procedures for drywell closure under emergency conditions are desirable, since offsite releases from a severe accident could have unacceptable consequences, as discussed in Section 6.9.1.

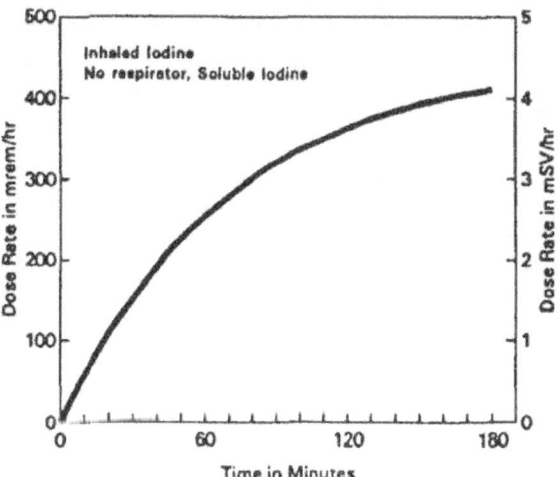

Figure 6.4 BWR Equivalent Whole-Body Dose (Inhaled Iodine)

6.9.5 Findings

- The estimated dose from a core melt 2 days after shutdown with an open containment is roughly 80,000 rem (thyroid) (800 Sv) and 200 rem (2 Sv) (whole-body) at a 1-mile distance from the plant. A closed PWR containment with 24-hour holdup followed by design rate leakage reduces these to 0.2 rem (2 mSv) (thyroid) and 0.001 rem (0.01 mSv) (whole body).

- BWR secondary containments are anticipated to fail within a few minutes of initiation of bulk boiling if steam is released into the secondary containment. Boiling can begin half an hour after RHR loss if loss occurs days after shutdown.

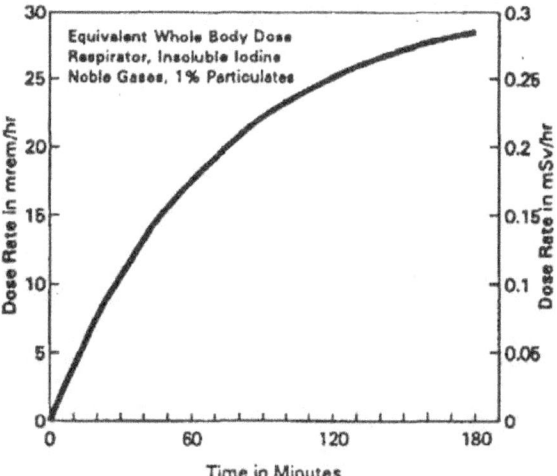

Figure 6.5 BWR Equivalent Whole-Body Dose (Noble Gases, Particulates)

- The plant visit program (see Chapter 3) found no BWRs for which primary containment closure was considered if RHR were lost. Existing secondary containments were judged to be of little use if the reactor vessel and primary containment were open.

- PWR licensee response was mixed concerning recommendations in GL 88–17 regarding containment closure. Some licensees have not fully evaluated attaining a no-gap equipment hatch closure. Closure techniques for other penetrations were sometimes poor. No licensee fully addressed the containment work environment if it planned to close the containment while steam was being released into the containment. Most closure procedures were weak and few had been rehearsed.

- Of the 107 plants surveyed, 52 required the use of ac power and/or compressed air to install the hatch. Five indicated that they had a procedure to close the hatch manually in the case of station blackout (SBO).

- Staff scoping analyses show that PWR containments probably require self-contained breathing apparatus within an hour of initiation of steam release into the containment due to the steam and temperature. (Localized heating and steam hazards were not considered.) Dose rates may not be serious if there are no fuel cladding leaks and if the licensee has significantly cleaned the water in the primary system, although breathing apparatus is likely to be needed. Airborne contaminants are of more concern with fuel leaks or contaminated primary water.

- Most primary containment concerns are eliminated if the containment is closed or if it is assured to be closed before the initiation of steam release from the RCS.

6.10 Fire Protection During Shutdown and Refueling

During shutdown and refueling outages, activities that take place in the plant may increase fire hazards in safety-related systems that are essential to the plant's capability to maintain core cooling. The plant technical specifications (TS) allow various safety systems to be taken out of service to facilitate system maintenance, inspection, and testing. In addition, during plant shutdown and refueling outages, major plant modifications are fabricated, installed, and tested. In support of these outage-related activities, increased transient combustibles (e.g., lubricating oils, cleaning solvents, paints, wood, plastics) and ignition sources (e.g., welding, cutting and grinding operations, and electrical hazards associated with temporary power) present additional fire risks to those plant systems maintaining shutdown cooling.

During plant shutdown, a postulated fire condition could potentially cause fire damage to the operable train or trains of residual heat removal capability. This fire damage could further complicate the plant's capability to remove decay heat.

In order to fully assess the fire risk during refueling conditions, the following action plan was implemented at a PWR and a BWR facility that the staff visited:

(1) Review the adequacy of current NRC fire protection guidance with respect to the protection of the systems necessary to perform the RHR function during shutdown and refueling modes of operation.

(2) Evaluate the fire protection requirements of Appendix R to 10 CFR Part 50 for cold-shutdown systems and determine if those requirements are adequate to

assure the availability of RHR capability under postulated fire conditions.

(3) Review administrative controls and methods for reducing fire hazards during shutdown and refueling modes of operation.

The results of this review and evaluation in each of the three areas are discussed next.

6.10.1 Adequacy of Current NRC Fire Protection Guidance for the Assurance of Residual Heat Removal Capability

The NRC fire-protection guidance (NUREG-0800, Standard Review Plan (SRP) Section 9.5.1) applied to ensure that an adequate level of fire protection exists, is a defense-in-depth approach. This approach is focused on the following programmatic areas:

- fire prevention through the use of administrative controls (e.g., good housekeeping practices, control of combustible materials, control and proper handling of flammable and combustible liquids, control of ignition sources)

- rapid fire detection through the use of early-warning fire-smoke-detection systems, fire suppression that occurs quickly through the application of fixed fire-extinguishing systems and manual fighting means, and limiting fire damage through the application of passive fire-protection features

- designing plant safety systems that provide for continued operation of essential plant systems necessary to shut down the reactor in those instances in which fire-prevention programs are not immediately effective in extinguishing the fire

The defense-in-depth concept, as it applies to fire protection, focuses on achieving and maintaining safe-shutdown conditions from a full-power condition. In addition, the SRP guidance given to licensees for conducting a fire hazard analysis specifies that the analysis should demonstrate that the plant will be able to perform safe-shutdown functions and minimize radioactive releases to the environment in the event that a fire occurs anyplace in the plant. The SRP guidance established for the performance of a fire hazard analysis does not address shutdown and refueling conditions, or the potential impact a fire may have on the plant's ability to remove decay heat and maintain reactor water temperature below saturation conditions.

The SRP establishes three levels of fire-damage limits for safety-related and safe-shutdown systems. The limits are established according to the safety function of the structure, system, or component. The following material summarizes the fire-damage limits: (1) one train of equipment necessary to achieve hot standby or shutdown (or both) from either the control room or emergency control stations must be maintained free from damage by a single fire, including an explosive fire; (2) both trains of equipment necessary to achieve cold shutdown may be limited so that at least one train can be repaired or made operable within 72 hours using onsite capability; and (3) both trains of systems necessary for mitigating the consequences following design-basis accidents may be damaged by a single fire. These damage limits are based on the assumption that full reactor power operation is the major limiting condition with respect to fire and its potential risk on reactor safety. The acceptable fire-damage threshold for RHR functions has not been established in the SRP with respect to the various shutdown and refueling modes of operation.

6.10.2 Evaluation of Requirements for Cold Shutdown

The Appendix R fire protection criteria for the protection of the safe-shutdown capability do not include those systems important to ensuring an adequate level of RHR during non-power modes of operation. Appendix R, Sections III.G and III.L, allow certain repairs to cold-shutdown components to restore system operability and the ability to achieve and maintain cold-shutdown conditions. This repair provision includes the decay heat removal functions of the RHR system. Appendix R requirements focus on full-power operation and address the impact a fire may have on the plant's ability to achieve and maintain safe-shutdown conditions.

During plant shutdown conditions in which the reactor head is removed, the RHR system and its associated support systems are performing the decay heat removal function (i.e., for PWR—component cooling water system, service water system, offsite/onsite ac/dc power train; for BWR—reactor building closed cooling water system, high-pressure service water system, offsite/onsite ac/dc power train). Depending on the specific mode of operation and the plant configuration (i.e., BWR/PWR—head off the vessel, water level at the vessel flange; PWR—head off in midloop operations), the plant technical specifications (TS) may require both trains or only one train of decay heat removal capability to be operable.

At one PWR facility visited, approximately 30 plant areas were associated directly with either the A- or B-train of decay heat removal. In 15 plant areas, both trains of RHR were present. This facility elected to comply with the Appendix R requirements by utilizing damage control/repair procedures. Under the Appendix R damage control/repair approach, a postulated fire during shutdown or refueling conditions in a plant area where both decay heat removal system trains are present, could cause fire damage to redundant trains resulting in a potential loss of

decay heat removal capability. By contrast, if the plant was at 100-percent power operations at the time of the fire, the plant could be held in hot standby until the necessary repair allowed under Appendix R could be made and subsequent cold shutdown could be achieved. For example, if the power cable to the RHR pump motor suffered fire damage, the plant maintenance staff estimated that it would take 16 hours to repair it and restore power to the pump. If this same postulated fire were to occur during shutdown or refueling, reactor coolant saturation conditions could potentially occur. As discussed in Section 6.6, there are several options available, depending on the plant configuration, for supplying water or providing limited RCS cooling. However, it should be noted that, without the performance of a detailed shutdown or refueling fire hazards analysis, the alternate RCS makeup and cooling options may have been affected by the same fire that caused the loss of decay heat removal.

During a visit to a BWR plant, it was determined that approximately seven areas of the reactor building and ten areas of the control building are associated with the decay heat removal function. Three areas in the reactor building and six areas in the control building contained both trains. In the areas containing both trains of decay heat removal, fire-protection features in accordance with Appendix R, Sections III.G and III.L, were provided. Since this plant's capability to achieve cold shutdown complies with Appendix R, Sections III.G and III.L, RHR fire-damage/control procedures were not required. However, by postulating a fire during shutdown and refueling conditions that required only one train of decay heat removal to be operable (the train provided with Appendix R fire protection is unavailable due to maintenance), in a plant area where the unprotected train is present, damage could be sustained to the operable train resulting in a total loss of decay heat removal capability. Under these conditions, RCS heatup to saturation could occur. There are several options available, depending on plant configurations, for supplying water to the RCS. These options include CRD pumps, standby liquid control system from test tank, condensate pumps, condensate or demineralized water via hoses from the service box on the fuel floor, core spray from the torus or condensate storage tank, refueling water transfer pump, high-pressure service water system, and makeup to reactor cavity skimmer surge tank and overflow into the reactor cavity. Alternate decay heat removal can be accomplished via the reactor cleanup or the fuel pool cooling systems. It should be noted that without the performance of a detailed shutdown and refueling fire analysis, the alternate RCS makeup and cooling options may not be available. The equipment or components (or both) associated with these options may be affected by the same fire that causes the loss of decay heat removal.

6.10.3 Review of Plant Controls for Fire Prevention

The staff reviewed fire-prevention administrative and control procedures associated with the control of transient combustibles and ignition sources, and the establishment of compensatory measures for fire-protection impairments. The fire-prevention administrative control measures are applicable to both power operation and shutdown conditions. It was noted that in order to support certain work activities (e.g., welding and cutting) associated with maintenance or modifications, a temporary fire prevention administrative-control procedure was changed. For example, a fire watch may be assigned to more than one welding or cutting operation, or increased combustible loading above that analyzed for full-power conditions may be introduced into safety-related areas to support maintenance operation. Fire-prevention administrative-control procedures did not provide enhanced controls or compensatory measures during shutdown conditions in those plant areas critical to supporting RCS makeup or decay heat removal.

During the PWR and BWR plant visits, when a plant walkdown was performed in areas that were associated with decay heat removal, an increase in fire hazards was noted. These fire hazards included temporary electrical and test wiring, increased transient combustibles (e.g., wood scaffolding, plastic sheeting and containers, lube oil, cleaning solvents, paper products, rubber products, and more), and increased welding and cutting activities. In addition, the staff noted that fire-protection personnel at the site had not increased their inspections. The staffing level is limited and fire-prevention inspections are restricted due to the increased paper work generated by activities associated with maintenance and modifications during an outage.

The lack of increased fire-prevention/protection activities commensurate with the increased maintenance and modification activities during plant shutdown and refueling is reflected by the increased frequency of fires. At the two facilities visited, a review of fire reports for an 18-month operating period showed that three fires occurred at the PWR and four fires at the BWR facility. Six of the seven fires at these facilities occurred during refueling outages.

6.10.4 Summary of Findings

- A postulated fire could potentially damage the operable train or trains of decay heat removal systems during shutdown conditions. In addition, plant configurations can further complicate the plant's ability to remove decay heat.

- Increased transient combustibles and ignition sources during outage activities present additional

fire risks to their minimum level of TS systems required to maintain shutdown cooling.

- SRP guidance established for the performance of a fire-hazard analysis does not address shutdown and refueling conditions and the potential impact a fire may have on the plant's ability to maintain core cooling.

- 10 CFR Part 50, Appendix R, fire-protection criteria for the protection of the safe-shutdown capability do not include those systems important to assuring an adequate level of decay heat removal during non-power modes of operation.

- Fire-prevention administrative-control procedures did not provide enhanced controls or compensatory measures during shutdown conditions in those plant areas critical to supporting RCS makeup or decay heat removal.

- The staffing level at the site for fire prevention is limited and inspection activities are restricted because so much paper work was generated by activities associated with maintenance and modifications during an outage.

- A majority of the fires at the facilities occurred during refueling outages.

6.11 Fuel Handling and Heavy Loads

Mishaps in handling fuels and heavy loads during the refueling process can occur and have a potential for

- causing an array of new or spent fuel to become critical

- damage to fuel assemblies which causes release of radioactivity

- overheating of spent fuel pool which damages fuel cladding

6.11.1 Fuel Handling

In order to minimize fuel-handling mishaps, the fuel-handling equipment is designed and built in accordance with specified standards to prevent dropping fuel. In addition, fuel-handling equipment is also tested before the fuel-handling process to assure its proper operation. Design guidelines for such equipment include the provision of high-temperature alarms and high-radiation alarms, should fuel damage or failures be imminent.

Criticality involved in the movement of a single fuel assembly is extremely unlikely with the greatest potential occurring in the case of misplacement of an element in the core or spent fuel pool. Proper planning and particular attention to details during the fuel-handling process can minimize the probability of mistakes. In BWRs, the potential for criticality during refueling is minimized by starting the process with the mode switch in the refueling or shutdown position and with all rods in. In PWRs, the boron concentration in the reactor coolant and refueling canal is kept at a level sufficient to assure a k_{eff} equal to or less than 0.95 or, as an alternative, the boron concentration is kept equal to or greater than 1850 ppm. In addition, licensees are required to analyze the worst case of fuel mislocation and provide assurance that the concomitant fuel damage does not cause offsite doses in excess of specified criteria.

The licensee is also required to analyze the condition for an uncontrolled control rod assembly (a bank for a PWR and a single rod for a BWR) withdrawal at subcritical or low-power condition, and to provide assurance that certain preset criteria, which includes thermal margin limits, fuel centerline temperatures, and uniform cladding strain for BWRs, are not exceeded.

Release of radioactivity from a spent fuel element may be caused by mechanical damage, such as dropping or striking it against some object. Dropping is minimized by proper design of handling equipment in accordance with specified criteria. Nevertheless, equipment has failed and fuel elements have been damaged. In order to minimize the radiation dosage as a result of such mishaps, all spent fuel must be moved under water during the refueling process. Current STS for both PWRs and BWRs require that a specified level of water must be maintained above the reactor vessel head and spent fuel storage pools during refueling. This level of water is capable of acting as shielding for the handling of spent fuel and for absorption of the radioactivity that could be released should a spent fuel element be damaged. In addition, the fuel handling equipment is tested before being used in order to avoid using faulty equipment, and to assure load handling limitations as required by TS.

For PWRs, TS require that penetrations in the containment building be closed or be capable of being closed by an operable automatic valve on a high-radiation signal in the containment, before initiating the refueling process. For BWRs, TS require that the integrity of the fuel-handling building be assured before handling irradiated fuel.

As a final protection against the potential excessive radiation doses resulting from a fuel-handling accident, the licensee must provide an analysis of the radiological consequences of a fuel-handling accident to assure that results will conform to applicable dose limitations.

Spent fuel in the spent fuel pool is kept cool by a spent fuel pool cooling system. TS for PWRs and BWRs require that such a system be operable in order to keep spent fuel cooled. TS also require that the water level in the spent fuel pools and temperatures be maintained to minimize dose levels during fuel handling. Spent fuel pool cooling systems are analyzed to ensure that proper spent fuel pool coolant temperatures are maintained at all times of storage of spent fuel so as to prevent overheating of the stored fuel.

6.11.2 Heavy Load Handling

In cases where access to the reactor core is required, it is necessary to remove the internal components. In doing so, the fuel elements could be damaged should a heavy load be dropped, resulting in the release of radioactive elements from damaged fuel. Relocation of damaged fuel into a critical mass is also of concern. Similar circumstances could occur upon lifting a heavy load over spent fuel elements stored temporarily in the containment or in the spent fuel storage pool.

Any heavy load carried over redundant equipment used for removal of decay heat has a potential for damaging or destroying this equipment or other equipment involved in shutdown. Damage, in such case, is limited by following safe load paths or by minimizing the potential for damage, as noted below.

Risk associated with heavy loads can be minimized as outlined in NUREG-0612, "Control of Heavy Loads at Nuclear Power Plants," (1) by making the potential for a load drop extremely small, by utilizing a single-failure-proof lifting system in accordance with NUREG-0612, or (2) by evaluating a potential load drop accident and taking actions to ensure that damage is so limited that

- Coolant lost can be replaced by normal makeup sources.

- The capability for systems to maintain safe shutdown is not lost.

In order to minimize the potential for a drop of a heavy load, licensees were required to (1) develop procedures for handling heavy loads, (2) train and qualify crane operators, (3) design special lifting devices in accordance with specified criteria, (4) design other lifting devices (other than "special") in accordance with specified criteria, (5) inspect, test, and maintain cranes in accordance with specified guidelines, (6) have cranes designed in accordance with specified criteria, and (7) follow safe load paths.

Three potential hazards regarding the handling of heavy loads are (1) damage to surroundings in the improper design or use of handling equipment so as to permit swinging or rotating of the load on breaking of one holding line; (2) improper handling of the internals of the Mark I BWRs and, by reference, of the internals of any reactor so as to damage the vessel, the core, or other safety-related equipment; and (3) dropping of loads placed on the edge of the spent fuel pool.

In each NRC regional office, a representative was contacted in an effort to determine whether problems had been observed in these areas. Only item 3 (i.e., dropping of loads from the edge of the fuel pool) was mentioned to be of concern, but not considered to be a significant shutdown risk issue.

There appears to be no special generic problem regarding handling heavy loads. Heavy reactor internals can be handled safely by adhering to the guidelines in NUREG-0612. The problem of load swing or rotation can be avoided by proper handling. Since the staff has not identified such an event, it concludes that load-handling procedures are being successfully employed in the field.

6.12 Onsite Emergency Planning

The staff's technical evaluation of shutdown and low-power operation shows that event sequences with potential offsite consequences can occur during cold-shutdown and refueling conditions. The plant configuration during shutdown and refueling conditions is significantly different from that during power operation. As a result, the sequence of events and the operator's ability to detect and respond to an event and mitigate its consequences may vary significantly during shutdown and refueling conditions. Therefore, the need for an operator to respond appropriately to an incident, including emergency classifications and notifications of offsite officials, still exists during cold-shutdown and refueling conditions.

6.12.1 Classification of Emergencies

Guidance for classifying emergencies at nuclear plants during power operation is found in Appendix 1 to NUREG-0654 (FEMA-REP-1), Revision 1, entitled "Criteria for Preparation and Evaluation of Radiological Emergency Response Plans and Preparedness in Support of Nuclear Power Plants." This guidance does not explicitly address the different modes of nuclear power plant operation. It is generally recognized, however, that the initiating conditions established in Appendix 1 to NUREG-0654 apply as a whole to a nuclear plant during its power operation and hot-shutdown modes. Some, but not all, of the initiating conditions in NUREG-0654 may apply to a nuclear plant during cold-shutdown and refueling conditions.

Because initiating conditions contained in Appendix 1 to NUREG-0654 were not intended to be directly and fully applicable to shutdown and refueling conditions and their unique characteristics, their use by the licensees has resulted in inconsistencies and oftentimes excess conservatism in the classification of emergencies during shutdown or refueling conditions. For example, the loss of vital ac power and RHR at Vogtle Unit 1 in March 1990 was classified as a Site Area Emergency by the licensee, but might have been classified as an Alert by a different licensee. In an event at Oyster Creek in March 1991, an Alert was declared when it was determined that both sources of onsite ac power were unavailable. However, offsite ac power was available at the time and the refueling cavity was flooded with water.

NUMARC has developed a method for defining emergency action levels which is referenced in NUMARC/NESP-007, Revision 1. Although the NUMARC approach is not considered complete in that regard. NRC will continue to work with NUMARC to issue the final guidance that will help licensees to identify initiating conditions and develop associated emergency action levels for shutdown and refueling conditions with a revised NUREG-0654. In the meantime, the staff will develop interim guidance for emergency classification during shutdown and refueling conditions. The interim is discussed in Chapter 7.

6.12.2 Protection of Plant Workers

NRC regulations in 10 CFR 50.47(b)(10) require that a range of protective actions be developed for emergency workers and the public. In meeting this requirement, as stated in Criterion J of NUREG-0654, the NRC expects each licensee to (1) evacuate nonessential personnel in the event of a Site Area or General Emergency and (2) account for onsite personnel within 30 minutes of the declaration of an emergency. During outage periods, hundreds of additional workers may be on site for maintenance, construction, and repairs. In addition to the presence of large numbers of workers on site during an outage, many unusual activities will be taking place and normally available equipment and instrumentation may not be available. These conditions, common during shutdown and refueling outages, can place an additional burden on the emergency response capability at the time of an accident. Emergency plans and procedures must address the evacuation and accountability of the large number of nonessential personnel on site should an accident occur during plant shutdown or refueling.

7 POTENTIAL INDUSTRY ACTIONS

7.1 Introduction and Perspective

The comprehensive technical evaluation of shutdown and low-power operations described in the previous chapters, included observations and inspections at a number of plants, analysis of operating experience, deterministic safety analysis, and insights from probabilistic risk assessments. From this evaluation, the staff has concluded that although public health and safety have been adequately protected during the period that plants have been in shutdown conditions, safety could be improved substantially and such improvement appears to be warranted for the following reasons:

(1) Significant precursor events involving loss of the decay heat removal (DHR) capability continue to occur despite NRC efforts to resolve the problem.

(2) Accident sequences during shutdown that are as rapid and severe as those that occur during power operation should be addressed with commensurate requirements. This is supported by the staff's engineering analysis of accidents during shutdown conditions documented herein.

(3) There is a significant lack of controls, including regulatory controls, that in the past allowed plants to enter circumstances likely to challenge safety functions with minimal mitigation equipment available and containment integrity not established.

There have been only a very limited number of probabilistic risk assessment (PRA) studies covering shutdown conditions, and those studies contain considerable uncertainty. The uncertainty is due largely to the predominant role played by operators and other licensee staff in shutdown events and recovery from them. Human reliability is difficult to quantify, especially under unfamiliar conditions which are often not covered in training or procedures. The collection of PRA studies discussed in Chapter 4 does give some insight into the plausible range of shutdown risks for the spectrum of current plants. The mean core-damage frequency (CDF) for shutdown events appears to be in the range of 3×10^{-5} to 7×10^{-6} per reactor-year. Although detailed uncertainty analysis is not available for most of the shutdown PRAs, some insight can be gained by examining the uncertainty analysis in NUREG-1150 where the ranges of CDF (5th and 95th percentiles) are approximately one order of magnitude. From this limited information, we conclude that a reasonable estimate of the range of CDF is 1×10^{-4} to 1×10^{-6} per reactor-year.) The risk to public health appears to be dominated by core damage in combination with an open or partially open containment. This would indicate that an improvement in CDF of about one order of magnitude is

warranted if it can be gained at a reasonable cost. In addition, an improvement in the likelihood of containment isolation when needed appears appropriate. As part of the regulatory analysis, the staff is presently determining the potential benefits and costs of all potential new requirements to the extent practical.

7.2 Previous Actions

Over the past 12 years, the NRC has issued eight generic letters related to shutdown and low-power operations. These generic communications present a chronology of events and actions requested by the NRC to preclude or mitigate events that could affect the nuclear power plant during low-power and shutdown operations. Generic Letter (GL) 88-17, "Loss of Decay Heat Removal," is the most comprehensive and most widely applicable of the generic letters. It specifically addresses shutdown concerns and is the most recent generic letter to make recommendations about low-power and shutdown operations. GL 88-17 made recommendations for operating pressurized-water reactors (PWRs) with reduced inventory; the recommendations concerned areas of instrumentation, administrative controls, operator procedures, and operator training.

Licensees have implemented GL 88-17 to varying degrees of effectiveness and completeness. All licensees operating PWRs have improved reduced inventory operation. Some licensees exceeded the GL 88-17 recommendations; others responded minimally. In GL 88-17, the staff limits its discussion to operation of PWRs during reduced inventory. The staff's positions given below expands coverage to conditions other than reduced inventory in PWRs and to some conditions in BWRs (boiling-water reactors).

The staff recognizes that industry has addressed shutdown and low-power operations with programs that include workshops, Institute of Nuclear Power Operations (INPO) inspections, Electric Power Research Institute (EPRI) support, as well as enhanced training and procedures. One activity (a formal initiative proposed by the Nuclear Management and Resources Council (NUMARC)) has produced a set of guidelines for the utilities to use for self-assessment of shutdown operations (NUMARC 91-06). This high-level guidance addresses many of the areas in outage planning that need to be improved. Detailed guidance on developing an outage planning program is outside the scope of the NUMARC effort; and, consequently, such important areas as fire protection and operability of mitigative equipment are not treated explicitly. The staff finds that NUMARC 91-06 represents a significant and constructive step, effects of which have already been realized by many utilities that used the guidance in recent outages; however, to address all of the issues relating to safe operation during outages requires more than the guidance of NUMARC 91-06.

7.3 Improvements in Shutdown and Low-Power Operations

The evaluation described in the preceding chapters indicates that additional requirements governing shutdown operation in the following areas are warranted:

- outage planning and control
- fire protection
- technical specifications
- instrumentation

7.3.1 Outage Planning, Outage Control, and Fire Protection

The technical findings in the previous chapters show that a more safety-oriented approach to planning and controlling outage activities will reduce risk during shutdown by reducing the incidence of precursor events and improving defense in depth. Such an approach should include (1) a comprehensive program for planning and controlling outage activities, including fire protection, and (2) limiting conditions for operation (LCOs), controlled through the plant technical specifications, for plant equipment needed to ensure key safety functions are available. It is better to use technical specifications rather than administrative controls to control the availability of safety-related equipment, because operators are already trained and accustomed to operating the facility within the clear limits set by technical specifications. In addition, the technical specifications establish clear and enforceable regulatory requirements.

Elements for an Outage Program

It is the staff's position that a complete program for planning and controlling outages in a safe manner would include those elements listed below:

(1) clearly defined and documented safety principles for outage planning and control

(2) clearly defined organizational roles and responsibilities

(3) controlled procedure defining the outage planning process

(4) early planning for all outages

(5) strong technical input based on safety analysis, risk insights, and defense in depth

(6) independent safety review of the outage plan and subsequent modifications

(7) planning and controls that (a) maximize the availability of existing instrumentation used to monitor temperature, pressure, and water level in the reactor vessel and (b) give accurate guidelines for operations when existing temperature indications may not accurately represent core conditions

(8) controlled information system to provide critical safety parameters and equipment status on a real-time basis during the outage

(9) contingency plans and bases, including those necessary to ensure that effective decay heat removal (DHR) during cold shutdown and refueling conditions can be maintained in the event of a fire in any plant area

(10) realistic consideration of staffing needs and personnel capabilities with emphasis on control room staff

(11) training

(12) feedback of shutdown experience into the planning process

Because the role of outage planning and control is central to shutdown safety, some regulatory controls to ensure adequacy and continued implementation at all plants may be appropriate. Controls could be imposed through a new regulation governing outages, or by including the requirements in the administrative controls section of the existing technical specifications. In either case, such a requirement would call for a program for planning and control of outages that

(1) includes the 12 elements listed above

(2) is documented in a controlled procedure subject to inspection by the NRC

(3) is subject to revision with the approval of an onsite safety review organization

7.3.2 Technical Specifications for Control of Safety-Related Equipment

Findings in previous chapters lead the staff to conclude that current standard technical specifications (STS) do not reflect the risk significance of many reactor coolant system configurations used during cold-shutdown and refueling operations. This is particularly true of technical specifications (TS) for PWRs. The staff also notes that some older plants do not have even basic TS covering residual heat removal (RHR) and electrical systems. TS are important because they establish the minimum safety standards during various operational conditions, and licensees carefully track them as a way of ensuring compliance with other regulatory requirements. The staff is considering the following changes to the current STS for BWRs:

(1) The specification in the STS for ac power sources during shutdown (i.e., "AC Sources—Shutdown") should be modified to require that redundant onsite emergency ac sources be operable during cold shutdown and refueling when the water level is less than [23] feet ([7] m) above the reactor pressure vessel flange and fuel is in the vessel. Redundancy is not required when the water level in the refueling cavity equals or exceeds [23] feet ([7] m) above the reactor pressure vessel flange because the passive cooling capability in the refueling cavity allows sufficient time to restore a DHR loop or establish an alternate method of cooling. This change ensures that the capability to remove decay heat will not be lost under such conditions as a loss of offsite power and a single failure of one onsite ac source.

(2) New specifications should be added to the STS to require operability of the plant service water system (standby service water system for BWR/6) and ultimate heat sink during Modes 4 and 5.

The staff is considering the following changes to the current STS for PWRs:

(1) The technical specifications for "RCS Loops-Mode 5, Loops Filled," and "RCS Loops—Mode 5, Loops Not Filled" should be combined into one specification for "RCS Loops—Mode 5." In addition, Action statements should be added requiring that (a) containment integrity be established if one required DHR loop becomes inoperable and cannot be returned to service in 8 hours, (b) an alternate method of DHR be established if both required DHR loops are inoperable, and (c) containment integrity be established if both required DHR loops become inoperable, and containment integrity has not been established by a separate specification on containment integrity discussed in item 4 below.

The requirements to establish an alternative method of DHR and achieve containment integrity, when one or more DHR loops becomes inoperable, are designed to ensure defense in depth when normal cooling systems become unavailable. In most cases, the emergency core cooling system will be available to serve as backup to cool the core and to act as a first line of defense when normal systems are unavailable.

(2) A technical specification for the emergency core cooling system during shutdown (i.e., "ECCS—Shutdown") should be added to require two trains of high-pressure injection (HPI) during cold shutdown and refueling with water level in the refueling cavity less than [23] feet ([7] m) above the reactor pressure vessel flange. The applicability for the STS on "ECCS—Operating" should be extended to Mode 4. Also, the applicability of the STS covering the refueling water storage tank (RWST) should be extended to Modes 5 and 6 when the water level in the refueling cavity is less than 23 [feet] ([7] m) above the reactor pressure vessel flange and when the cavity is not being flooded with water from the RWST.

(3) Because of the change to the technical specification for "ECCS—Operating," the specification for "Low Temperature Overpressure Protection" should be modified to require either a larger size vent to mitigate the effects of higher mass addition from a second HPI train, or specific controls to isolate HPI trains during shutdown and refueling (i.e., keep HPI discharge valves closed and tagged and keep pumps in pull-to-lock).

(4) Studies in previous chapters indicate that shutdown risk is highest when decay heat is high and the reactor coolant system (RCS) is in a condition of reduced inventory. In light of this, a new specification should be added to the STS to require containment integrity under these conditions. This specification should require containment integrity to be maintained during Mode 5, should natural circulation cooling not be available, and Mode 6, should the water level in the refueling cavity be less than [23] feet ([7] m) above the reactor vessel flange. Containment integrity in these modes should not be required after the core decay heat has been reduced below a plant-specific value so that the containment can be closed manually before boiling occurs in the RCS, assuming a loss of the offsite electrical power system and the unavailability of the onsite ac power system.

Some licensees may have alternate emergency ac sources available to them, as defined in 10 CFR 50.2, or portable power supplies that could be used to assist in manually closing the containment. This equipment should be credited in estimates of the times to manually close the containment, if these power supplies are ensured to be available through the outage plan.

The staff believes that such an LCO would give licensees considerable flexibility, through planning and good engineering, to minimize the need for containment integrity during normal activities while in Mode 5 or 6.

(5) New specifications should be added to the STS to require redundant systems for (a) component

cooling water and (b) service water and operability of the ultimate heat sink during cold shutdown and refueling when the water level in the refueling cavity is less than [23] feet ([7] m) above the reactor pressure vessel flange. Redundancy is not necessary during refueling when the water level in the refueling cavity equals or exceeds [23] feet ([7] m) above the reactor pressure vessel flange because the passive cooling capability in the refueling cavity allows sufficient time to restore a DHR loop or establish an alternate method of cooling.

(6) The specification in the STS for ac power sources during shutdown (i.e., "AC Sources—Shutdown") should be modified to require redundant onsite emergency ac sources to be operable during cold shutdown and refueling when the water level in the refueling cavity is less than [23] feet ([7] m) above the reactor pressure vessel flange. Redundancy is not necessary when the water level in the refueling cavity equals or exceeds [23] feet ([7] m) above the reactor pressure vessel flange because the passive cooling capability in the refueling cavity allows sufficient time to restore a DHR loop or establish an alternate method of cooling. The specification should also require that if ac sources become inoperable and cannot be returned to service within 8 hours, the equipment supported by those sources must be declared inoperable. Alternate sources of ac power that may be available at some sites during shutdown operations may be credited under some conditions. However, the staff would consider plant-specific technical specifications that factor in such sources on a case-by-case basis.

(7) Action statements should be added to the specification in the STS on "DHR and Coolant Circulation—Low Water Level" to require that, with one DHR loop inoperable, the water level in the refueling cavity be raised to least 23 [feet] ([7] m) above the reactor vessel flange or that containment integrity be established if the loop cannot be returned to service within 8 hours. If both required DHR loops are inoperable, an alternate method of DHR must be established. Containment integrity must be established before boiling occurs in the reactor coolant system if both required DHR loops become inoperable and containment integrity has not already been established.

The requirements to establish an alternate method of DHR and achieve containment integrity when one or more DHR loops becomes inoperable are designed to ensure defense in depth when normal cooling systems become unavailable. In most cases, the emergency core cooling system will be available to serve as backup for core cooling and to act as a

first line of defense when normal systems are unavailable.

(8) Action statements should be added to the technical specification for "DHR and Coolant Circulation—High Water Level" to require that with no DHR loops operable or in operation, containment integrity should be established within 8 hours.

7.3.3 Water-Level Instrumentation in PWRs

PWR licensees have added level instrumentation to cover shutdown operation in response to GL 88-17 and, in PWRs, level indications have generally improved in the last 3 years. However, events in PWRs continue to occur (e.g., Prairie Island, 1992) in which existing methods for monitoring water level have failed to adequately indicate a level too low to support DHR pump operation. Consequently, the staff is considering a potential requirement for licensees of PWRs to install an additional means of accurately monitoring water level in the RCS during mid-loop operation. This additional instrumentation should not be affected by errors induced in the other level measurements by changes in pressure in the RCS or connected systems. Normally, ultrasonic devices or other such local measurements as pressure differential across the hot leg will be needed for meeting this criterion. The installed instrumentation should include visual and audible indications in the control room to alert operators to an inappropriate condition. The instrumentation should be placed in operation before the plant enters a reduced inventory condition.

7.4 Other Actions Considered

In the course of its evaluation of key issues of shutdown risk, the staff considered one additional potential industry action but chose not to pursue it. This was to issue a supplement to Generic Letter 88-20, "Individual Plant Examination (IPE) for Severe Accident Vulnerabilities," asking licensees to include shutdown and low-power conditions in their IPEs. The reasons for not pursuing this action at this time are discussed below.

The intent of the IPE program is to identify plant-specific deficiencies mostly involving hardware and not directly or effectively handled in the licensing process. The shutdown risk program is aimed at resolving generic issues associated with operations during shutdown and low-power operation and this can be done most effectively with generic requirements. However, not having a shutdown IPE program at this time doesn't mean that the staff wishes to discourage licensees from applying risk-based methods to understand the implications of shutdown activities or to help in planning outages. Another important reason for not recommending an IPE for shutdown and low-power conditions at this time is that IPE is dependent on a well-developed and understood PRA methodology, and this does not currently exist for shutdown and low-power conditions. The current IPE program follows more than a decade of experience with PRAs for power operation. The NRC Office of Regulatory Research expects to complete its PRAs for shutdown and low-power conditions in FY94.

7.5 Conclusions

The staff is considering a number of potential requirements that can address shutdown and low-power issues effectively. These are appropriate for the following reasons:

(1) The potential requirements reflect the traditional NRC safety philosophy of defense in depth in that they address (a) prevention of well-understood and credible challenges to safety functions through improvements in outage planning and fire protection; (b) mitigation of challenges by redundant protection systems, well-founded procedures, and training, and through improved technical specifications and contingency plans; (c) availability and reliability of containment through improved technical specifications and response procedures; and (d) emergency preparedness through improved contingency plans.

(2) The potential requirements are aimed directly at problems that have been observed repeatedly in operating experience, e.g., loss of DHR, loss of ac power, loss of RCS inventory, fires, personnel errors, poor procedures, poor planning, and lack of training.

In accordance with the backfit rule, 10 CFR 50.109, the staff is currently performing a formal regulatory analysis to determine if the potential requirements discussed above will yield a substantial improvement in safety and are cost effective. A final decision will be made on the need for new requirements and the form they should take after the Commission, the Advisory Committee on Reactor Safeguards, and the Committee To Review Generic Requirements have reviewed the issues.

8 POTENTIAL NRC STAFF ACTIONS

As discussed in Chapter 1 and in SECY 91-283, the staff has evaluated a number of key issues regarding shutdown risk, and additional technical issues. By means of this review, the staff has identified potential actions that can improve the following NRC programs as they relate to shutdown and low-power operations: the licensing reviews for advanced reactor design, the inspection program, the operator licensing program, and the program for analysis and evaluation of operational data. In addition, probabilistic risk assessment (PRA) studies of shutdown and low-power conditions at Surry and Grand Gulf will continue.

From a more general viewpoint, the staff has reconfirmed that nuclear reactor safety is the product of prudent, thoughtful, and vigilant efforts of the NRC and the licensees and not the result of "inherently safe" design or "inherently safe" conditions. The current areas of weakness in shutdown operations stem primarily from the false premise that "shut down" means "safe." The primary staff action must be a recognition of this fact and a resolve not to substitute complacency for appropriate safety programs.

8.1 Advanced Light-Water-Reactor Reviews

Insights from the shutdown operation work are being factored into future light-water reactor design reviews. Staff members conducting these reviews have periodically met with staff personnel working on shutdown issues since Generic Letter (GL) 88-17 was issued, and appropriate concerns have been addressed both in meetings with industry and in questions asked of the industry. As previously discussed, the April 30 through May 2, 1991, inter-office meeting on shutdown/low-power issues identified issues and topics for further consideration. These insights were incorporated into questions asked of industry representatives working on future light-water-reactor designs. This work is continuing. For example, several meetings have been held with General Electric on shutdown issues for the advanced boiling-water reactors and with ABB Combustion Engineering on the System 80+ design. The findings and conclusions reached in this report will be reviewed for applicability to these designs, and appropriate initiatives will be taken to ensure their adequate consideration.

8.2 Proposed Changes to the Inspection Program

The staff reviewed the current NRC inspection program to determine how the program could be expanded to better address shutdown issues. As a preliminary result, the staff has developed a temporary instruction (TI) for the conduct of a shutdown risk and outage management team inspection. The staff has conducted five pilot inspections at Oconee Unit 2, Indian Point Unit 3, Diablo Canyon Unit 1, Prairie Island Units 1 and 2, and Cooper Station to fully develop the TI. The staff is continuing to assess the need for this type of team inspection. Shutdown risk and outage management are being evaluated as a potential topic for the mandatory team inspection program. The results of these activities, upon their completion, will be presented to the Commission with recommendations.

8.2.1 Assessment of the Inspection Program

The staff examined its current inspection program to see if it needed to be improved. As described in NRC Inspection Manual Chapter 2515, "Light-Water Reactor Inspection Program—Operations Phase," the inspection program comprises three major program elements:

(1) core inspections

(2) discretionary inspections (which include regional initiative inspections, reactive inspections, and team inspections)

(3) area-of-emphasis inspections (which include generic area team inspections and safety-issues inspections)

Issues of shutdown and low-power risk are addressed to varying degrees in each of the three major program elements in Manual Chapter 2515. Recent changes to core-inspection procedures have added emphasis to monitoring operations during shutdown conditions. A number of reactive inspections, including several augmented inspection teams and one incident investigation team inspection, have been conducted in response to shutdown events. Safety-issues inspections have also been conducted to verify implementation of recommended actions and program enhancements required by GL 88-17. A recently issued TI (2515/113) also addressed inspection of licensee activities and administrative controls for reliable decay heat removal during outages. These inspections have succeeded in directing attention to issues of shutdown and low-power risk. However, recurring problems in the area of outage management indicate a possible need for an increased inspection emphasis in this area.

8.2.2 Team Inspection

A generic area team inspection could focus NRC and industry attention on the area of outage management, should the Commission desire such emphasis. The inspection would assess the effectiveness of licensee programs for planning and conducting plant outage activities. As currently envisioned, the inspection would consist of a minimum of 2 weeks of onsite inspection by a team of five inspectors (including the site resident inspector). These inspections would be scheduled to coincide with the conduct of a planned outage. The first week of the inspection would coincide with an outage planning and the second with the outage period. Emphasis would be placed on the following areas:

- management involvement and oversight of outage planning and implementation

- the relationships among significant work activities and the availability of electrical power supplies, decay heat removal systems, inventory control systems, and containment capability

- the procedures and training related to controlling plant configuration during shutdown conditions

- areas in which operations, maintenance, and other plant support personnel work together and channels of communications between them

- supervision of work activities and control of changes to the outage schedule

- assurance of component and system restoration before plant restart

- operator response procedures, contingency plans, and training for mitigation of events involving loss of decay heat removal capability, loss of reactor coolant system inventory, and loss of electrical power sources during shutdown conditions

- the operator's ability to monitor plant status in order to detect and classify an emergency

The pilot inspections have identified some plant deficiencies with respect to control of outage activities. In particular, at Oconee Unit 2, a required nuclear instrument reliability check had not been performed during fuel movement, and at Indian Point Unit 3, a commitment in response to Generic Letter 88-17 concerning residual heat removal (RHR) pump motor current indication had not been satisfactorily completed. The pilot inspections found that licensees were beginning to implement the NUMARC 91-06 guidelines. The outage planning process had been modified to address such areas as assessing the risks associated with planned outage activities and scheduling outage activities to minimize overall plant risk.

8.2.3 Inspection of the Use of Freeze Seals

Loss of freeze seals used in pipe connections on the bottom of the reactor vessel head in boiling-water reactors (BWRs) could cause a rapid loss of reactor coolant and a potential for core uncovery. Other concerns with the use of freeze seals are discussed in Section 6.6.1. The staff concluded that freeze seals should be treated as plant modifications and, therefore, should be evaluated in accordance with requirements of 10 CFR 50.59. Consequently, the staff intends to revise the NRC Inspection Manual to include guidance on application of 10 CFR 50.59 to freeze-seal operations to ensure that proper safety evaluation is performed and unreviewed safety questions are identified. This revision will be evaluated to determine if it constitutes a backfit (i.e., change of a staff position) and will be presented to the Committee To Review Generic Requirements for review.

8.3 Operator Licensing Program

The staff recognizes that operators who have proper knowledge and understanding of risks associated with shutdown can greatly reduce risk associated with outage activities. This knowledge and understanding can be increased through training programs that give more emphasis to shutdown operations. The staff also recognizes that although the current Nuclear Regulatory Commission (NRC) Examiner Standards (NUREG-1021) allow for coverage of shutdown operations, the standards do not specify what constitutes an acceptable level of coverage. Consequently, the staff revised the current NRC Examiner Standards. The standards for the initial examination have been revised to strengthen reference information and ensure that at least one job performance measure related to shutdown and low-power operations was evaluated. The standards for requalification examinations have been revised to (1) place more emphasis on shutdown operations and (2) review the licensee's requalification exam test outline for coverage of shutdown and low-power operations, consistent with the licensee's Job Task Analysis and Operating Procedures. These changes are incorporated in NUREG-1021, Revision 7.

8.4 Analysis and Evaluation of Operational Data

The Office for Analysis and Evaluation of Operational Data (AEOD) is performing an analysis of shutdown and low-power (SD/LP) operational data. This special study (similar in approach to the Reactor Scram Study, NUREG-1275, Volume 5) is an assessment of existing SD/LP operating performance and is designed to provide a baseline and the process for trending future performance. The analysis will identify industry-wide indicators and provide a means of assessing trends for SD/LP issues. The study will also evaluate the availability of operational data for effective assessment of SD/LP performance trends.

SD/LP event reporting practices were also reviewed. Although sufficient information is available to analyze SD/LP operating performance, weaknesses were identified. AEOD will incorporate the results of this review in a revision to NUREG-1022, "Event Reporting Systems, 10 CFR 50.72 and 50.73," as appropriate.

8.5 PRA Studies

The Office of Nuclear Regulatory Research (RES) conducted PRA investigations of shutdown and low-power operations at Surry and Grand Gulf in several stages. Quantitative findings in the form of point estimates for the level 1 internal events have been completed. Results for the seismic and internal fire and flooding analyses will follow later in 1993.

An uncertainty analysis and a comprehensive report covering this stage will be completed by the end of 1993. This will include a conventional PRA for the complete set of level 1 sequences, to be followed by a more comprehensive analysis using state-of-the art methods.

RES has also performed abridged level 2 and 3 analysis for Surry and Grand Gulf for specific plant operating states (i.e., specific portions of the overall low-power and shutdown mode). The results of these analyses indicated that the consequences of a core-meltdown accident could be significant, at least for the specific operating states stud-

ied, in which many of the containment barriers were unavailable.

8.6 Emergency Planning

The Nuclear Management and Resources Council (NUMARC) has developed a system similar to that in NUREG-0654 for classifying abnormal occurrences at nuclear power plants. The NUMARC methodology is documented in NUMARC/NESP-007, Revision 2, "Methodology for Development of Emergency Action Levels." In developing this system, NUMARC has recognized that initiating conditions are more accurately defined when the plant's mode of operation is considered. In the NUMARC methodology, initiating conditions are dependent on the reactor mode of operation. The NRC staff endorsed the NUMARC methodology in Regulatory Guide 1.101, Revision 3, issued in August 1992.

Although the NUMARC methodology includes initiating conditions for nuclear plants during shutdown and refueling, it is not considered complete in that regard. NUMARC intends to complete its analysis of the findings of the NRC's shutdown and low-power evaluation and to develop an industry position on possible further guidance. The NRC staff will coordinate its efforts with NUMARC to develop and issue guidance that will help licensees identify initiating conditions* and to develop associated emergency action levels for the shutdown and refueling conditions.

---

*The initiating conditions listed in Appendix 1 to NUREG-0654 are used by each licensee to develop emergency action levels based on site-specific measurable/observable plant indicators.

9 REFERENCES

The references listed here were used to varying degrees in conducting this evaluation and preparing this report. They are arranged by issuing body or author.

American National Standards Institute

ANSI/ANS–3.5–1985   "Nuclear Power Plant Simulators for Use in Operator Training."

Battelle Columbus Laboratories   "Development of Guidelines for Use of Ice Plugs and Hydrostatic Testing," November 1982.

Brookhaven National Laboratory   "PWR Low Power and Shutdown Accident Frequencies Program—Phase 1: Coarse Screening Analysis," Draft Letter Report, November 13, 1991.

"PWR Low Power and Shutdown Accident Frequencies Program—Phase 2: Internal Events," Draft Letter Report, August 31, 1992.

"Fire Risk Analysis of POS 6 and 10 Surry Low Power and Shutdown Project" Draft Letter Report, January 8, 1993.

Electric Power Research Institute

EPRI NP–6384–D   "Freeze Sealing (Plugging) of Piping," February 1989.

Idaho National Engineering Laboratory

EGG–EAST–9337, Rev. 1   "Thermal-Hydraulic Processes Involved in Loss of Residual Heat Removal During Reduced Inventory Operation," February 1991.

Jacobson, S.   "Some Local Dilution Transients in a Pressurized Water Reactor," Thesis No. 171, LIU–TEK–LK–1989;11, Linkoping University, Sweden.

Nuclear Management and Resources Council

NUMARC 91–06   "Guidelines for Industry Actions To Assess Shutdown Management," December 1991.

NUMARC/NESP–007, Rev. 1   "Methodology for Development of Emergency Action Levels," 1991.

Nuclear Regulatory Commission

AEOD Special Report   "Review of Operating Events Occurring During Hot and Cold Shutdown and Refueling," December 4, 1990.

Bulletin 80–12   "Decay Heat Removal System Operability," May 1980.

CN 87–22   "NRC Inspection Manual."

Generic Letter 80–53   "Transmittal of Revised Technical Specifications for Decay Heat Removal Systems at PWRs," June 1980.

Generic Letter 81–21   "Natural Circulation Cooldown," May 1981.

Generic Letter 82–12   "Nuclear Power Plant Staff Working Hours," June 15, 1982.

Generic Letter 85–05   "Inadvertent Boron Dilution Events," January 1985.

Generic Letter 86–09   "Technical Resolution of Generic Issue B–59, (n–1) Loop Operation in BWRs and PWRs," March 1986.

Generic Letter 87–12	"Loss of Residual Heat Removal (RHR) While the Reactor Coolant System (RCS) Is Partially Filled," July 1987.
Generic Letter 88–17	"Loss of Decay Heat Removal," October 17, 1988.
Generic Letter 88–20	"Initiation of the Individual Plant Examination for Severe Accident Vulnerabilities," August 1989.
Generic Letter 90–06	"Resolution of Generic Issues 70, "Power-Operated Relief Valve and Block Valve Reliability," and 94 "Additional Low-Temperature Overpressure Protection for Pressurized Water Reactors" [pursuant to 10 CFR 50.54(f)]," June 1990.
Information Notice 91–36	"Nuclear Plant Staff Working Hours," June 1991.
Information Notice 91–41	"Potential Problems With the Use of Freeze Seals," June 27, 1991.
Information Notice 91–54	"Foreign Experience Regarding Boron Dilution," September 1991.
Memorandum	From J.M. Taylor to the Commissioners, "Staff Plan for Evaluating Safety Risks During Shut-down and Low Power Operations," October 22, 1990. (Available in the Public Document Room attached to minutes of public meeting with NUMARC, November 7, 1990.)
NUREG–0612	"Control of Heavy Loads at Nuclear Power Plants," July 1980.
NUREG–0654, Rev. 1	"Criteria for Preparation and Evaluation of Radiological Emergency Response Plans and Preparedness in Support of Nuclear Power Plants" (FEMA-REP-1), November 1980.
NUREG–0800	"Standard Review Plan for the Review of Safety Analysis Reports for Nuclear Power Plants," June 1987.
NUREG–1021, Rev. 6	"Operator Licensing Examiner Standards," 1990.
NUREG–1022	"Event Reporting System, 10 CFR 50.72 and 50.73," September 1991.
NUREG–1122	"Knowledges and Abilities Catalog for Nuclear Power Plant Operators: Pressurized Water Reactors," December 1989.
NUREG–1123	"Knowledges and Abilities Catalog for Nuclear Power Plant Operators: Boiling Water Reactors," December 1989.
NUREG–1150	"Severe Accident Risks: An Assessment for Five U.S. Nuclear Power Plants," December 1990.
NUREG–1269	"Loss of Residual Heat Removal System," June 1987.
NUREG–1275	"Reactor Scram Study," March 1989, Addendum August 1989.
NUREG–1410	"Loss of Vital AC Power and the Residual Heat Removal System During Mid-loop Operations at Vogtle Unit 1 on March 20, 1990," June 1990.
NUREG/BR–0150, Rev. 1	"RTM 91: Response Technical Manual," Volumes 1 and 2, AEOD, April 1991.
NUREG/CR–4674	"Precursors to Potential Severe Core Damage Accidents: 1990, A Status Report," Oak Ridge National Laboratory, Volume 14, September 1991.
NUREG/CR–5015	"Improved Reliability of Residual Heat Removal Capability in PWRs as Related to Resolution of Generic Issue 99," Brookhaven National Laboratory, May 1988.

NUREG/CR-5819	"Probability and Consequences of Rapid Boron Dilution in a PWR: A Scoping Study," Brookhaven National Laboratory, March 1992.
NUREG/CR-5820	"Consequences of Loss of the RHR System in Pressurized Water Reactors," Idaho National Engineering Laboratory, May 1992.
Regulatory Guide 1.101	"Emergency Planning and Preparedness for Nuclear Power Reactors," Rev. 3, August 1992.
Regulatory Guide 1.149	"Nuclear Power Plant Simulation Facilities for Use in Operator License Examinations," April 1987.
SECY 91-283	"Evaluation of Shutdown and Low Power Risk Issues," James M. Taylor, Executive Director for Operations, to The Commissioners, September 9, 1991.
Site Access Training Manual	Compiled by NRC Technical Training Center, June 1991.

Nuclear Safety Analysis Center EPRI/NSAC)

NSAC-52	"Residual Heat Removal Experience and Safety Analysis, Pressurized Water Reactors," January 1983.
NSAC-83	"Brunswick Decay Heat Removal Probabilistic Safety Study," Final Report, October 1985.
NSAC-84	"Zion Nuclear Plant Residual Heat Removal PRA," July 1985.
NSAC-125	"Industry Guidelines for 10 CFR 50.59 Safety Evaluations," June 1989.
Salah, S., et al.	"Three Dimensional Kinetics Analysis of an Asymmetric Boron Dilution in a PWR Core," *Trans. ANS*, 15:2 (1972)
Sandia National Laboratories	"BWR Low Power and Shutdown Accident Sequence Frequencies Project—Phase 1: Coarse Screening Analysis," Draft Letter Report, November 23, 1991.

APPENDIX A

Cold Shutdown Event Analyses

This appendix documents the precursor analyses of ten cold shutdown events. This documentation includes (1) a description of the event, (2) additional event-related information, (3) a description of the model developed to estimate a conditional core damage probability for the event, and (4) analysis results. A table of contents, Table A.1, follows.

Table A.1. Index of cold shutdown analyses

ACCIDENT SEQUENCE PRECURSOR PROGRAM COLD SHUTDOWN EVENT ANALYSIS

LER No.:   271/89-013 R1
Event Description: Reactor cavity draindown
Date of Event: March 9, 1989
Plant: Vermont Yankee Nuclear Power Station

Summary

Vermont Yankee maintenance personnel established a reactor cavity leak path on March 9, 1989 when they performed required post-maintenance testing on a residual heat removal/shutdown cooling (RHR/SDC) suction valve. Operators took more than 47 min to determine the flow path for the resultant drain-down which transferred about 10,300 gal of water to the suppression pool. The leak path was isolated in two min once the source of the leak was discovered. The conditional core damage probability estimated for this event is less than 1×10^{-6}.

Event Description

On March 4, 1989, Vermont Yankee placed the "B" loop of RHR into SDC and took the "A" loop out of service for maintenance. Five days later the "A" and "C" RHR pump motors were racked out for maintenance. System logic, in effect at that time, opened the min-flow valve for these pumps. About 15 h later, electrical maintenance personnel racked out the "A" and "C" SDC suction valves. Following the repair work on the valves, the technicians manually stroked open the valves as required by procedure. This established a leak path for the reactor cavity. Personnel working on the refuel floor notified the control room operators within five min that they had noticed an 18" drop in the reactor cavity water level. The operators thought this was due to the refilling of the recently opened portion of the "A" RHR loop. However, 15 min later the refuel floor personnel reported another 18" drop in level. The refuel floor was evacuated, as a result, and the operators began to search for the leakage path. Refuel floor personnel reported additional level decreases at 15 min intervals. Successive level drops of 24" and 60", following the first two 18" drops, were noted before the control room operators discovered the leak path. An operator was sent to close the manual isolation valve in the minimum flow line which isolated the leak path.

It should be pointed out, RHR SDC was never lost and the reported total level drop was 120" while the measured drop was 72". The latter measurement was based on the inventory increase in the suppression chamber. Further, this event could only have occurred with vessel head removed.

Fig. 1 is a simplified drawing of the RHR system.

Additional Event-Related Information

Initial water level was about 290" above top of active fuel (TAF), this corresponds to 13" below the reactor vessel flange. Primary containment isolation system automatic initiations occur at 127" above TAF. Specifically, a reactor scram and the automatic isolation of the RHR SDC from the reactor recirculation system. Emergency core cooling system (ECCS) initiation occurs at 82.5" above TAF. Upon ECCS initiation, RHR automatically lines up for low-pressure coolant injection (LPCI) mode. That is, valves line up for pump suction on the suppression chamber, SDC isolation, and test return isolation.

ASP Modeling Assumptions and Approach

Analysis for this event was developed based on procedures (e.g. Procedure OP 2124, Rev. 20, Issued October 13, 1988) in effect at Vermont Yankee at the time of the event, the Plant Technical Specifications, and the Final Safety Analysis Report. While the following assumptions are specific to Vermont Yankee, they are applicable to most contemporary boiling water reactors (BWRs).

a. Core damage end state. Core damage is defined for the purpose of this analysis as reduction in reactor pressure vessel (RPV) level above TAF or unavailability of suppression pool cooling in the long term. With respect to RPV inventory, this definition may be conservative, since steam cooling may limit clad temperature increase in some situations. However, choice of TAF as the damage criterion allows the use of simplified calculations to estimate the time to an unacceptable end state.

b. Prolonged maintenance on an RHR train (as in this event) is only likely with the reactor head removed. Therefore, only this head state was considered in the analysis. If the head is removed, then any makeup source greater than ~200 gpm, combined with boiling in the RPV, will provide adequate core cooling.

c. Four makeup sources were available during this event: low-pressure coolant injection (LPCI), core spray, control rod drive (CRD) flow and the feedwater/condensate system. Use of any other source of makeup is considered a recovery action.

The event tree model for the event is shown in Fig. 2. If the loss of inventory is corrected before RPV isolation (as was the case during the event), then RHR cooling is maintained. Once RPV level decreases to the RHR SDC isolation setpoint (127" TAF) and either of the RHR suction line isolation valves close, normal shutdown cooling is lost. In this case, RPV makeup using LPCI, core spray, CRD flow or the condensate/feedwater system will provide continued core cooling. LPCI and core spray will automatically initiate once RPV inventory drops to the ECCS initiation setpoint (82.5"), if not initiated manually before this point. If RHR SDC isolation fails, then one LPCI or core spray pump will provide sufficient makeup to offset the loss through the open min-flow valve.

The following branches are included on the event tree:

<u>Inventory Loss Terminated Before RHR ISO</u>. Operator action to identify and isolate the inventory loss prior to the RHR SDC isolation setpoint will prevent loss of SDC. Based on simplifying assumptions, it is estimated that the vessel level would reach the RHR SDC isolation setpoint in approximately 1.8 h.

Assuming (1) an exponential repair model, (2) that the observed time to detect and isolate is the median time for such actions, and (3) that no isolation was possible during the first 20 min (to account for required response and diagnosis), a probability of 0.1 is estimated for failing to isolate the inventory loss prior to reaching the RHR SDC isolation setpoint.

<u>Inventory Loss Terminated by RHR ISO</u>. Closure of either of the SDC suction isolation valves will isolate the RHR system and terminated the loss of inventory. Based on the failure probabilities used in the ASP program, a probability of failing to isolate RHR of 1×10^{-3} is estimated. If one division were unavailable, a probability of 1×10^{-2} would be estimated.

<u>LPCI Flow Available</u>. On Vermont Yankee, one or more RHR/LPCI pumps take a suction from the suppression pool (i.e. torus) and discharge to the core via the reactor recirculation loops. RHR/LPCI consists of two redundant trains, each of which includes two parallel RHR/LPCI pumps, one suction valve (open when a train is aligned for LPCI, closed when aligned for SDC), and one discharge (RPV injection) valve (closed when a train is aligned for LPCI, open when aligned for SDC).

In this event, the pumps in one of the two trains were unavailable because of maintenance. Injection success for the operating train requires the suppression pool suction valve for the operating RHR pump to open. If this valve fails to open, the non-operating pump must start and its suction valve must open. Based on probability values used in the ASP program, a LPCI failure probability of 3.7×10^{-4} is estimated. It was assumed that normally-open valves and check valves do not contribute substantially to system unavailability.

<u>Core Spray Flow Available</u>. For Vermont Yankee, the core spray system consists of two trains. Each train includes one pump with a single, normally open motor-operated suction valve and a single normally-closed discharge (RPV injection) valve. The pump suction source is normally the suppression pool. Based on the probabilities used in the ASP program, a failure probability of 6.8×10^{-4} is estimated. If one division were unavailable, this probability would be 6.8×10^{-3}. It was assumed that normally-open valves and check valves do not contribute substantially to system unavailability.

<u>CRD Flow Available</u>. At cold shutdown pressures, one of two CRD pumps can provide makeup. Since one pump is typically running, the system will fail if that pump fails to run or if the other (standby) pump fails to start and run. Assuming a probability of 0.01 for failure of the standby CRD pump to start, and 3.0×10^{-5}/hr for failure of a pump to run, results in a estimated failure probability for CRD flow of 3.0×10^{-6}. In this estimate, a short-term, non-recovery likelihood of 0.34 was applied to the non-running pump failure-to-start probability, consistent with the approach used to estimate the failure probability for the core spray system. A mission time of 24 h was also

assumed.

If only one train is available (because of maintenance on the opposite division), then the CRD failure probability is estimated to be 7.2×10^{-4}.

Feedwater/Condensate Available While the feedwater or condensate pumps can provide more than adequate makeup, they are often unavailable during a refueling outage because of work on the secondary system; however, for this event, the feedwater/condensate system was available. A failure probability of 0.01 was assumed on this analysis.

For this event, substantial time existed to recover equipment failures. If RHR isolation was successful, more than 24 h would have been required before core uncovery. This long period of time is primarily due to the large volume of vessel inventory above the core and the relatively low decay heat load from the core. If RHR isolation failed, 1.4 h would have been required to reduce RPV level to TAF. These estimates are based on an initial water level 13' below the top of the vessel flange. Normally, with the head off, the reactor cavity would be flooded, which would add significant additional inventory.

Analysis Results

Based on the model described above, the conditional probability of severe core damage for this event is estimated to be less than 1.0×10^{-6}. This low value reflects the multiplicity of systems available to provide continued core cooling and the reactor vessel head status believed to be required before conditions which lead up to the event could have occurred.

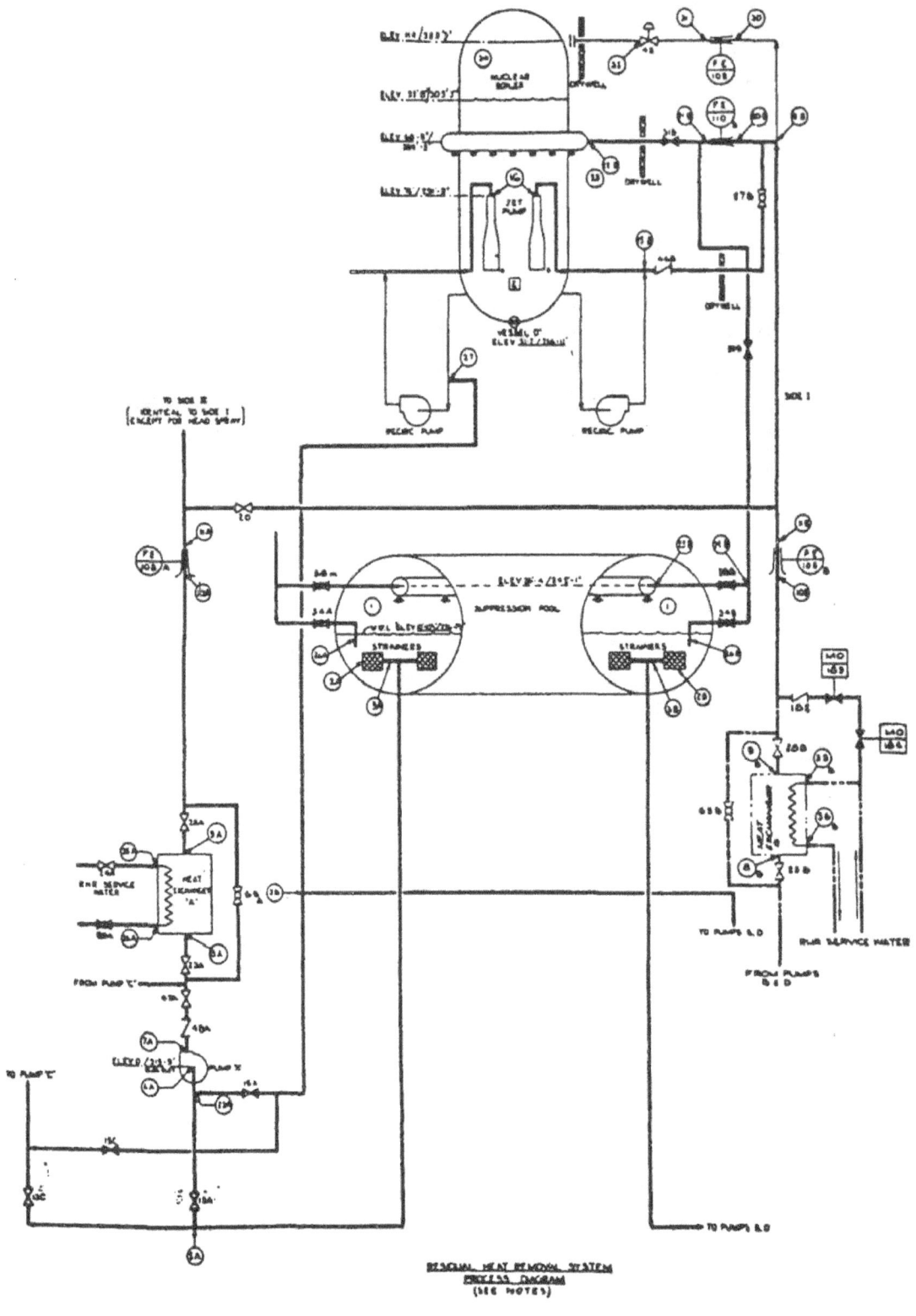

Fig. 1. Vermont Yankee RHR System

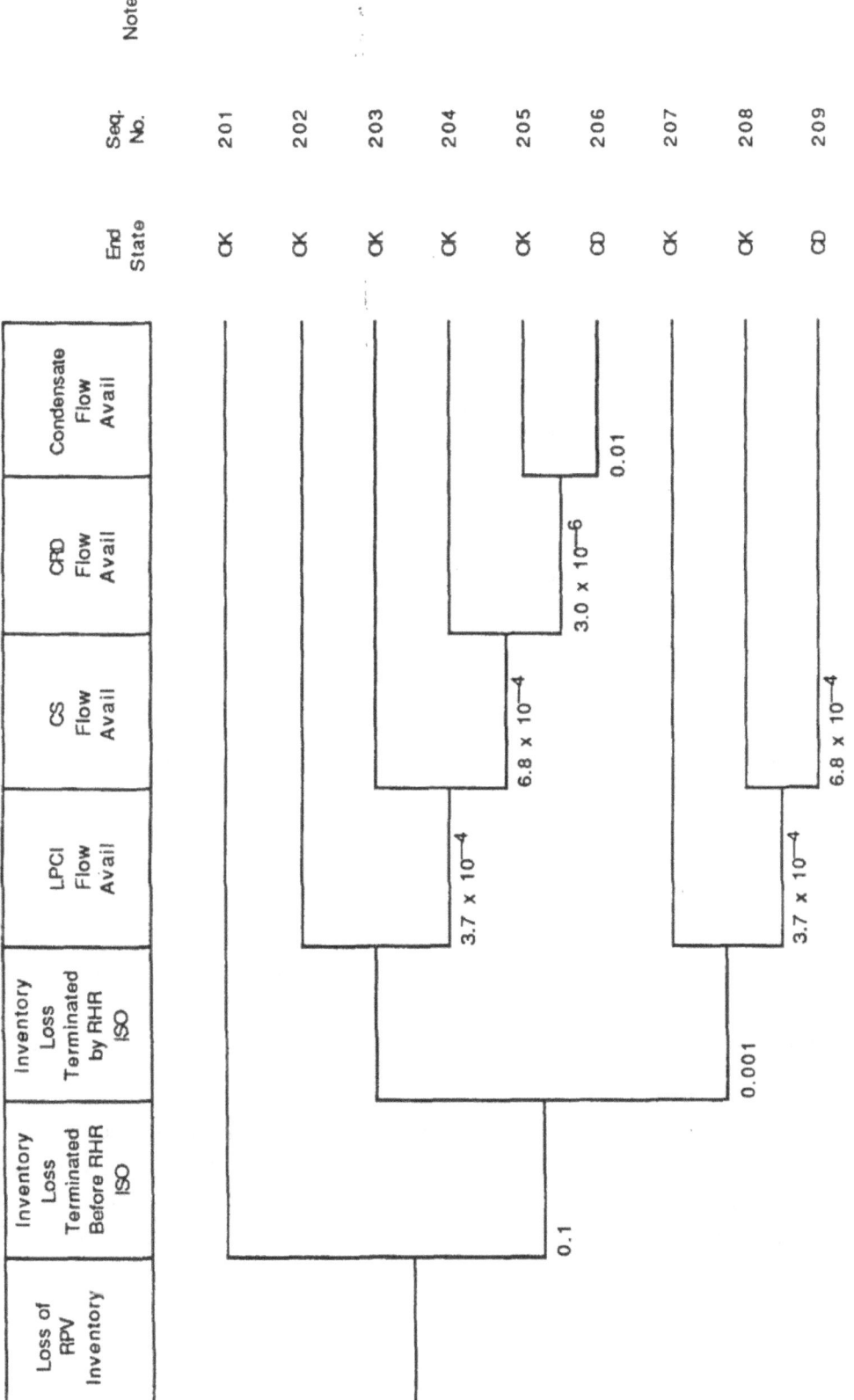

Fig.2. Event tree model for LER 271/89-013 R1

Notes: 1. Alternate sources of service water may also provide injection

ACCIDENT SEQUENCE PRECURSOR PROGRAM COLD SHUTDOWN EVENT ANALYSIS

LER No.: 285/90-006
Event Description: Loss of offsite power, diesel fails to load automatically.
Date of Event: February 26, 1990
Plant: Fort Calhoun

Summary

During a refueling outage, a spurious relay actuation resulted in isolation of offsite power supplies to Fort Calhoun. One diesel generator (DG) was out of service for maintenance, the other started but was prevented from connecting to its electrical bus by a shutdown cooling pump interlock. Operators identified and corrected the problem, and the DG was aligned to restore power to the plant. The conditional probability of core damage estimated for this event is 3.6×10^{-4}. The dominant sequence involves failure to recover AC power or provide alternate RCS makeup from the RWT prior to core uncovery. The calculated probability is strongly influenced by estimates of failing to recover AC power in the long term. These estimates involve substantial uncertainty, and hence the overall core damage probability estimated for the event also involves substantial uncertainty.

Event Description

On February 26, 1990, on the ninth day of a refueling outage and with the RCS partially filled (above mid-loop) to support control element assembly uncoupling, spurious actuation of a switchyard breaker backup trip relay opened circuit breakers supplying power to 4160 V buses 1A1, 1A2, 1A3, and 1A4 from the plant 22 kV system. Normal power supplies to ESF buses 1A3 and 1A4 are from the 161 kV system, but these supplies had been removed to support maintenance activities. Emergency power supplies are provided for buses 1A3 and 1A4. The emergency power source for bus 1A3, DG D1, was out of service for maintenance, so no emergency power was available to that bus. The backup power source for bus 1A4, DG D2, started but was prevented from energizing the bus by an interlock in a low-pressure safety injection (LPSI) pump circuit. This resulted in interruption of all AC power supplies to plant equipment.

Prior to the event, LPSI pump "B" had been placed in service for residual heat removal. The plant electrical system is designed such that, if a LPSI pump has been manually started and a subsequent loss of offsite power occurs, the LPSI pump breaker cannot be opened automatically and the DG output breaker for the affected train cannot be closed to feed its ESF bus. Thus, while DG D2 started correctly in response to the undervoltage condition on bus 1A4, the LPSI pump remained tied to the bus and the DG could not supply its loads.

Approximately one minute after the loss of offsite power (LOOP), plant operators opened the LPSI pump breaker and DG D2 energized bus 1A4. The pump was then returned to service for shutdown cooling. Thirteen minutes later, offsite power was restored to bus 1A3.

Event-Related Information

Current plant procedures (pp 5-6 of AOP-32, "Loss of 4160 Volt or 480 Volt Bus Power") address the need to manually trip an operating RHR pump breaker before attempting to power the bus from its DG. Note that Rev. 0 of this procedure was issued in February 1991. However, the operators were able to restore shutdown cooling within 44 seconds, which indicates knowledge of this design condition did exist.

ASP Modeling Approach and Assumptions

Of interest in this event is the ability of plant operators to determine the need to remove loads from a deenergized ESF bus before attempting to repower from the emergency DG. This requirement is currently proceduralized and operator actions during the actual event show that the operators did not experience difficulty in repowering the bus.

The probability value used in the ASP program for failure of a single DG to start and supply its loads is 0.05. The likelihood that operators would fail to open the LPSI pump breaker, allowing the DG to feed ESF loads, is considered to be small in comparison. Therefore, the interlock design feature was not separately modeled.

During shutdown and refueling operations, a loss of AC power will result in loss of shutdown cooling/decay heat removal. The amount of time that decay heat removal can be unavailable before core damage results is a function of a number of variables including core power history, time since shutdown, water level in vessel, heat sinks available, and refueling configuration (head off/on, cavity flooded/not flooded, etc.).

The most limiting case occurs during mid-loop operation (reactor coolant drained to level of main coolant nozzles) with a high decay heat load (see discussion of Vogtle event, NUREG-1410). With lesser decay heat loads and/or a larger volume of coolant in the reactor coolant system (RCS), additional time exists for recovery actions. The likelihood of success for such actions has not been well quantified to date. However, it is believed that the increased likelihood of success associated with the additional time available when the plant is not in mid-loop more than compensates for the higher fraction of time that the plant is in a non-mid-loop condition, and that the risk associated with mid-loop therefore dominates.

In this event, the LOOP occurred early in a refueling outage, when decay heat loads could be expected to be fairly large. One train of emergency power was out of service. Fort Calhoun was above mid-loop at the time of the event. However any of three states may be found nine days into a refueling outage: mid-loop, normal shutdown, or refueling (reactor head off and cavity filled). As discussed, the first case is believed to dominate risk.

The event was modeled as a loss of offsite power during mid-loop operation. The event tree model is shown in Fig. 1. Recovery of RHR is not specifically shown, but is assumed to occur within one-half hour of recovering power to the safety-related buses. This time period reflects the potential need to vent the RHR system if reactor vessel inventory is lost because of boiling. Note that use of gravity feed from the RWT for RCS makeup is not viable at Fort Calhoun because of the location of the tank, and hence is not addressed in the model.

Branch probabilities were estimated as follows:

1. RCS level (mid-loop). The likelihood of a LOOP during mid-loop operation is estimated to be 0.11, based on NUREG-1410 (pp 6-7). Assuming the occurance of a LOOP is independent of the shutdown RCS status, the likelihood of being in mid-loop, given a loss of offsite power occurs during shutdown, is 0.11.

2. Emergency power fails. One DG was unavailable prior to the event. Since operator action to trip the operating RHR pump (to allow DG load) is not believed to appreciably impact the overall emergency power reliability, a nominal DG failure probability of 0.05 was assigned to this branch.

3. Offsite power recovered prior to saturation. By interpolation of data from NUREG-1410, it was estimated that, in mid-loop operation, the RCS coolant inventory would have reached saturation temperature in approximately 1 h. Recovery of offsite power prior to this time was assumed to prevent core damage. A probability of not recovering offsite power within one hour of 0.25 was used in the analysis. This probability was estimated using the plant-centered LOOP recovery curves in NUREG-1032 by assuming (1) that the observed time to recover offsite power (14 min) represented the median of such recovery actions and (2) that the shape of the plant-centered non-recovery distributions were representative for this event.

4. AC power recovered prior to core uncovery. Recovery of offsite power or the faulted DG and successful restart of RHR (including any required venting) or provision of pressurized RCS makeup is assumed to prevent core damage. Assuming core uncovery would occur in about 3 h, a probability of failing to recover AC power by that time, given that it was not recovered at 1 h, of 0.26 is estimated.

Analysis Results

The estimated conditional core damage probability associated with the LOOP at shutdown, given that one emergency DG was unavailable, is 3.6E-04. This value is essentially unrelated to the "design feature" which prevented auto DG loading if an RHR pump was in operation. The conditional probability is strongly influenced by assumptions regarding operator actions to align emergency power. It is also influenced by the assumption that no procedural requirement exists to prevent one DG being removed from service for maintenance at the same time that the RCS inventory is reduced below normal levels.

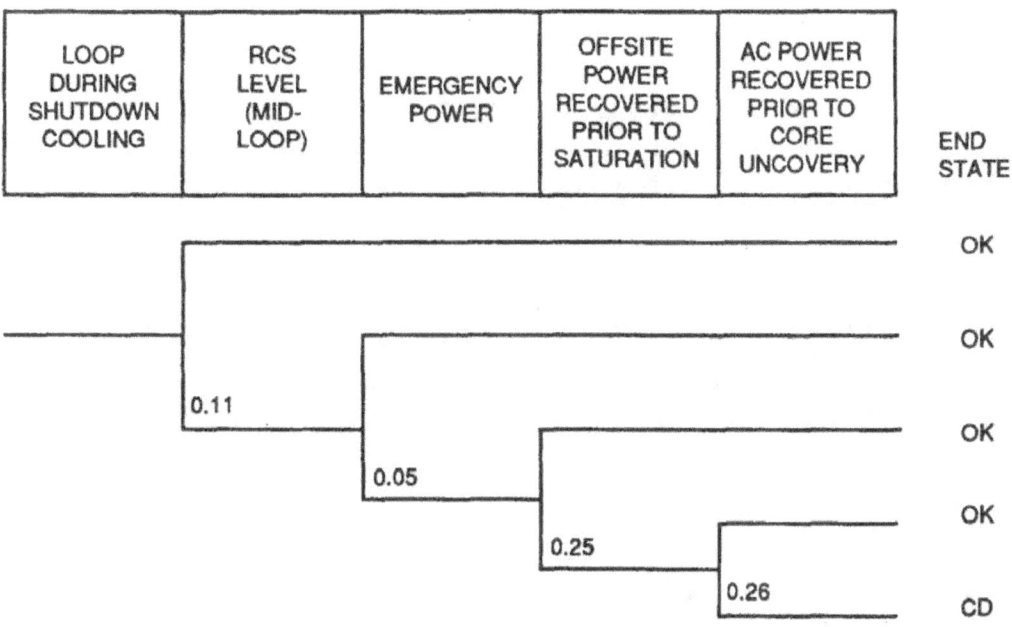

Fig. 1. Core Damage Event Tree for Loss of Offsite Power During Refueling
Outage at Fort Calhoun

ACCIDENT SEQUENCE PRECURSOR PROGRAM COLD SHUTDOWN EVENT ANALYSIS

LER No.:   287/88-005
Event Description: Errors during testing resulted in a 15 min loss of shutdown cooling during mid-loop operation
Date of Event: September 11, 1988
Plant: Oconee 3

Summary

A loss of AC power occurred at Oconee 3 while at mid-loop as a result of errors during emergency power switching logic circuit testing. This loss of power, which had to be recovered by local breaker closure, resulted in a 15 min loss of decay heat removal. The conditional probability of core damage estimated for the event is 1.7×10^{-6}. The dominant sequence involves failure to recover main feeder bus power from either of two offsite sources and failure to implement alternate reactor coolant system (RCS) makeup using the standby shutdown facility. Had this event occurred at a later time, when the current loss of the low pressure injection (LPI) system procedure was in effect, the conditional probability would be estimated to be below 1.0×10^{-6}. This is a result of the additional methods of decay heat removal specified in the current procedure.

Event Description

Oconee 3 was in cold shutdown with RCS in mid-loop. Test procedure PT/3/A/0610/01H, "Emergency Power Switching Logic Standby Breaker Closure Channel A & B," was started to test the circuitry for the emergency power switching logic. A decision was made to use the "Procedure for Removing From or Returning to Service 6900/4160/600 Volt Breakers," (R&R procedure) during the test. This decision was made since the breaker checklist, which confirms that groups of breakers are properly aligned, had already been completed in preparation for Unit 3 startup. The control room supervisor did not review the R&R procedure to identify any differences between it and the emergency power switching logic test procedure. In actuality, differences did exist and inapplicable sections of the R&R procedure should have been so marked by the control room supervisor.

During performance of the test, questions were raised by the non-licensed operator (NLO) responsible for aligning the breakers about an inconsistency between the two procedures regarding racking in breakers. The test procedure required this be done with the control power fuses removed to prevent spurious breaker trips when trip signals were generated, while the R&R procedure required control power fuses to be installed before breaker closure. This inconsistency was resolved by the control room supervisor, but inapplicable sections of the R&R procedure were still not marked.

Later in the test, the NLO originally responsible for aligning the breakers was reassigned to another task. A second NLO, who was now supporting the emergency power switching logic test, also questioned the inconsistency between the two procedures (he had been verbally informed the R&R procedure was being used after he had aligned breakers based only on the switching logic test procedure). The control room supervisor who had reviewed the two procedures was unavailable because of a meeting, and the unit supervisor instructed the NLO to restore the control power fuses in accordance with the R&R procedure.

Upon installation of the control power fuses, breaker 3B1T-1 tripped open and a loss of power occurred on Unit 3. At the time of the trip (0317), decay heat removal was being accomplished through the LPI system. RCS temperature was 90F. Upon the loss of power, the operating LPI pump was deenergized and decay heat removal capability was lost. Since the incore thermocouples had not been reconnected and the loss of power caused a failure of Reactor Vessel Level Transmitter 5, there were no available indications to determine the condition of the core. Even though the reactor protective system indications are battery-backed, these indications come from hot leg and cold leg resistence temperature detectors (RTDs), which were not available due to the system being open and due to the ongoing outage work.

The first method that was used in an attempt to restore power was to open the standby breakers and try to close breaker 3B1T-1 to provide power from the startup bus. This method was attempted since it was initially believed that 3B1T-1 tripped because the standby breakers were closed when the control power fuses were installed.

What actually tripped the breaker was a trip signal from a variable voltage transformer being used during the performance of the emergency power switching logic test. However, when the loss of power occurred, the variable voltage transformer also lost power. This resulted in a no-power-on-the-startup-bus-condition being sensed by the breaker, which prevented the breaker's closing. Operations personnel then racked the standby bus breakers into the closed position and energized the standby bus through those breakers.

When the standby bus was energized at 0332, the loss of power was terminated and the LPI pumps were restarted, and decay heat removal capability was again established. The core temperature was found the have risen approximately 15 degrees to approximately 105F. At 0355, an ALERT was declared on Unit 3 due to the "Loss of Functions to Maintain Plant Cold Shutdown" which occurred during the loss of power from 0317 to 0332. The ALERT was terminated at 0410.

Event-Related Information

At the time of the event, Unit 3 had completed refueling. The reactor vessel head was in place but not bolted, the RCS was depressurized, and RCS loops were drained to approximately 15 in above loop center line. One LPI pump was operating for decay heat removal, maintaining core coolant temperature at approximately 90F. The reactor building equipment hatch was open; therefore, containment was not closed at the time of the event. The reactor status was approximately 32 d after shutdown. When power was lost to the LPI pumps, decay heat removal was lost.

The subject event was analyzed by Duke Power, using actual plant conditions. Based on this analysis, the water in the vessel was expected to reach saturation 125 min after the loss of decay heat removal. Subsequent boiling would lead to core uncovering 10 h after saturation occurred.

In the Duke Power Company response to Generic Letter 87-12, a worst case scenario was analyzed for loss of decay heat removal while the RCS is depressurized. In this scenario, the RCS is depressurized and drained to 10 in above the loop center line elevation, the temperature initially at 100F, and the refueling canal drained. With a loss of decay heat removal occurring 72 h after shutdown, core uncovery was predicted to occur at 2 h and 41 min.

The "Loss of Low-Pressure Injection System" procedure (AP/3/A/1700/07) applicable at the time of the event specified the following if the RCS was opened and both LPI trains were inoperable: evacuate the reactor building and establish containment integrity, utilize one HPI pump with suction from the BWST to maintain RCS inventory (and RCS temperature <200F if thermocouples are available), and if the fuel transfer canal is full, use the spent fuel coolers to maintain RCS temperature. Use of gravity feed from the Boric Water Storage Tank (BWST) is not specified in the procedure in place at the time of the event.

The "Loss of Power" procedure (AP/3/A/1700/11) applicable at the time of the event specified reenergizing the main feeder from the startup source (transformer CT3), the Keowee hydro units (transformer CT4), or from the Lee gas turbines (transformer CT5). If none of these sources were available, operators were instructed to start the Standby Shutdown Facility (SSF) diesel and provide RCS makeup using the SSF RCS makeup pump or provide RCS makeup using HPI pump powered from the auxiliary service water pump switchgear (which is powered from standby bus 1). SSF RCS makeup is provided by a 26 gpm positive displacement pump. Based on simplified calculations and scaling of other analysis results, this pump can compensate for boil off at 22 d after shutdown (eight days after shutdown if the core is refueled).

A simplified diagram of the Oconee power system is shown in Fig. 1.

The current loss of power procedure is similar to the earlier procedure for actions applicable to this event, but with supplemental information added. The current loss of LPI system procedure has been expanded to include detailed instructions for establishing containment integrity and for providing RCS makeup using gravity feed from the BWST.

ASP Modeling Assumptions and Approach

The event has been modeled as a loss of decay heat removal during mid-loop as a result of the unexpected breaker trip and subsequent loss of power to the main feeder buses. All actions specified in the loss of LPI system procedure which existed at the time of the event required operable electrically-powered pumps. Since recovery of power to the main feeder buses would also recover power to the LPI pumps, alternate decay heat removal methods available once power was recovered were not included in the model. Instead, the event tree model considered two possible means of providing continued decay heat removal: restoring power to the main feeder buses by closing one of the breakers from a powered offsite source (transformer CT-3 and CT-5) or providing RCS makeup from the SSF RCS makeup pump.

An additional complication in the analysis is the short, 1-h battery lifetime identified for Oconee in the FSAR. Probabilistic risk assessments (PRAs) typically assume battery lifetime can be extended following a station blackout by shedding less important loads. In addition, battery lifetimes at cold shutdown are also expected to be greater than just after a trip from power (see ASP analysis of the March 20, 1990 event at Vogtle, documented in NUREG/CR-4674, Vol. 14, "Precursors to Potential Severe Core Damage Accidents: 1990, A Status Report"). It was assumed in this analysis that the battery lifetime would be greater than the time required to manually rack in the breakers and restore main feeder bus power.

The event tree model is shown in Fig. 2. Event tree branch probabilities were estimated as follows:

1. Main feeder bus recovered. Based on the time available to perform the proceduralized actions regarding recovery of main feeder bus power, only the likelihood of equipment (breaker) failure was considered when estimating this branch probability. Using a probability of 1×10^{-3} for failure of one of the breakers to close, and typical conditional probabilities of 0.1, 0.3 and 0.5 for failure of the second, third, and fourth breakers results in an estimated probability of 1.5×10^{-5} for failure to recover main feeder bus power from an offsite source.

2. SSF RCS makeup provided. Failure of this branch would occur if the SSF diesel or the SSF RCS makeup pump failed to start and run. A failure probability of 0.11 was employed, based on the analysis documented in the Oconee PRA (NSAC-60, Vol. 3, "Oconee PRA: A Probabilistic Risk Assessment of Oconee Unit 3").

Analysis Results

The conditional core damage probability estimated for this event is 1.7×10^{-6}. This low value reflects the fact that an alternate, proceduralized approach for decay heat removal was available, and that power for the LPI system could be easily recovered prior to battery depletion or core uncovery by manual operation of redundant breakers.

If this event occurred earlier in the refueling, when the small SSF RCS makeup pump could not make up for boil-off, a core damage probability of 1.5×10^{-5} would have been estimated. However, the decision which precipitated the event (use of the R&R procedure in conjunction with the emergency power switching logic test procedure) was made because the plant was near the end of the outage.

Had this event occurred at a later time, when the current loss of LPI system procedure was in effect, the conditional probability would be estimated to be below 1×10^{-6}. This is a result of the current requirement to use gravity feed from the BWST for RCS makeup.

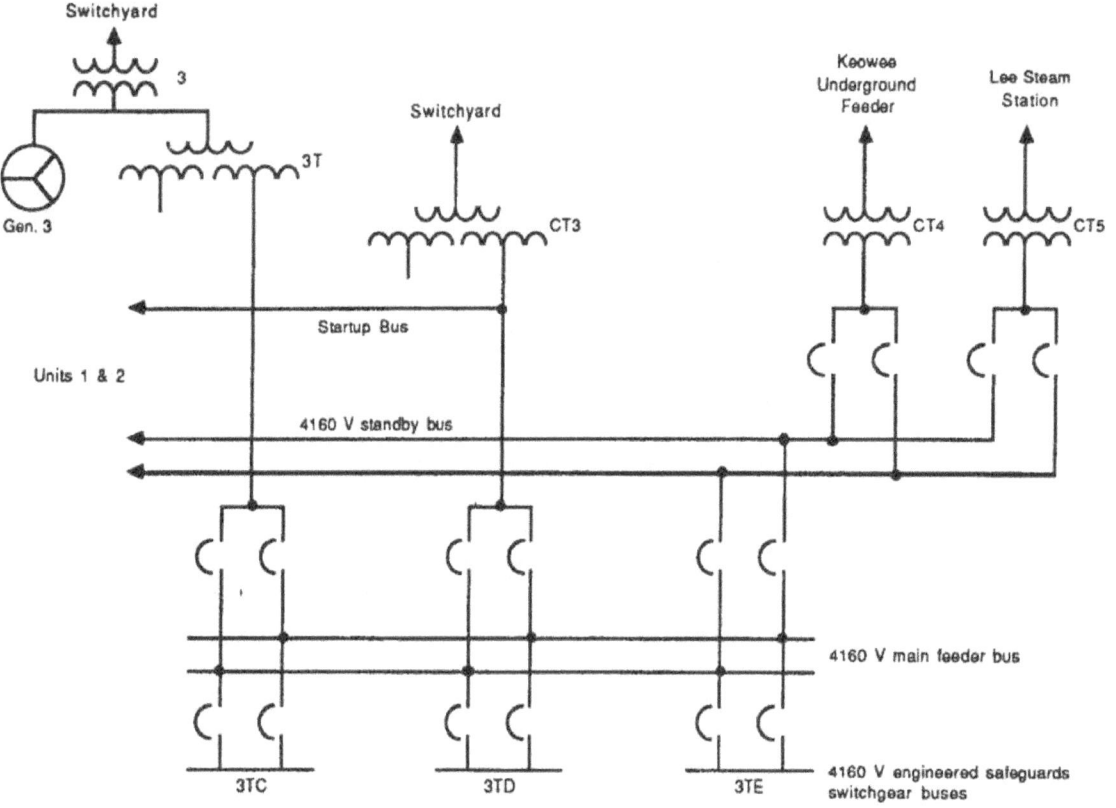

Fig. 1. One line diagram of the Oconee 3 power system

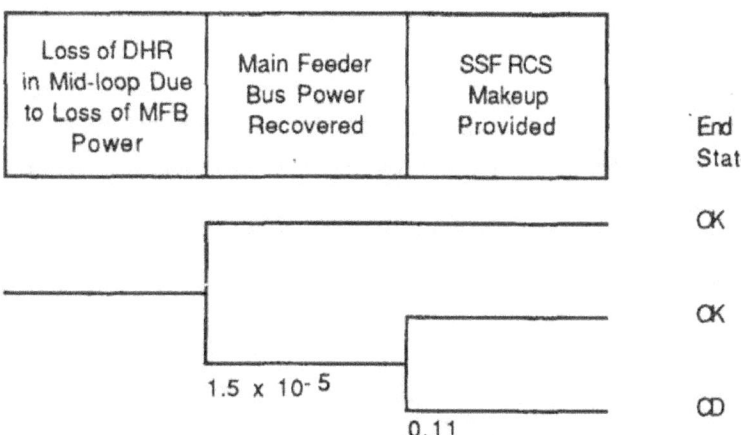

Fig. 2. Event tree model for LER 287/88-005

ACCIDENT SEQUENCE PRECURSOR PROGRAM COLD SHUTDOWN EVENT ANALYSIS

LER No: 302/86-003
Event Description: Loss of decay heat removal for 24 min due to pump shaft failure and redundant loop suction valve failure
Date of Event: February 2, 1986
Plant: Crystal River Unit 3

Summary

Crystal River Unit 3 was in cold shutdown when the "B" train of decay heat removal (DHR) was lost due to a pump shaft failure. The suction isolation valve for the "A" train DHR pump would not open on demand from the control room. An operator was sent to manually open the isolation valve. DHR capability was re-established approximately 24 minutes after the "A" train pump failed. Reactor coolant system (RCS) temperature rose to 131F from 98F during the period that DHR capability was unavailable.

Procedures identify 5 alternate means of providing DHR capability in addition to the "B" train of the DHR system. This event is estimated to have a probability of fuel damage of less than 1.0×10^{-6}.

Event Description

Crystal River Unit 3 was in cold shutdown and was performing repairs on a reactor coolant pump. The reactor coolant level was below the level of the reactor coolant pumps and the RCS was vented to atmosphere. Reactor vessel temperatures were being maintained at 98F by the "B" train of DHR. At 21:48, the "B" DHR pump, DHP-1B, tripped due to a motor overload caused by a failed pump shaft. Action was taken to place the "A" train in operation; however, the isolation valve (DHV-39) on the suction side of pump "B" failed to open on demand from the control room. Valve DHV-39 was manually opened and DHR was restored at 22:12. RCS temperature rose to 131F during the period that DHR capability was unavailable.

After repair of the damaged pumps, personnel observed movement of the pump and piping when water was being added to the system in order to fill this train of DHR. An examination revealed that several pipe restraints in the vicinity of the pump were loose or damaged.

Event Related Plant Information

The motor of DHP-1B overloaded and tripped as a result of a failed pump shaft. A failure analysis indicated that the failure occurred due to torsional fatigue induced by excessive shaft loading. The excessive shaft loads were most likely the result of pump air entrainment due to vortexing that occurred during operations at low RCS levels.

The failure of the suction isolation valve, DHV-39, to open on demand was a combination of several problems. Lubrication of the operator drive shaft and universal joints may have been inadequate. The operator torque switch setting was too low and the circuit breaker setpoint was too low for the motor load. Isolation valve DHV-39 was originally a manually operated valve. Its motor operator was installed in response to a NUREG-0578 item.

Crystal River 3 procedure AP-360, "Loss of Decay Heat Removal," has been substantially revised since 1986, when this event occurred. In the 1986 version, the operators are instructed to first start the alternate decay heat removal train, if available. If the alternate decay heat train cannot be started, then the procedure identified the use of OTSG cooling (if available) or SFC system, which can be tied to the DHR system on Crystal River. The use of high-pressure injection (HPI), low-pressure injection (LPI) or gravity feed from the borated water storage tank (BWST) to provide makeup to delay core uncovery is not identified in the 1986 procedure.

The current procedure has been updated to identify the following additional actions to maintain core cooling: flooding the fuel transfer canal, use of core flood tank inventory, and low- or high-pressure injection with suction from the BWST or reactor building sump.

Internals vent valves are installed in the core support shield on Crystal River 3 to prevent a pressure imbalance which might interfere with core cooling following a cold leg break. These valves are closed during normal operation, but in the event of a break in the cold leg, open and vent steam generated in the core directly to the break. During the 1986 loss of DHR, the RCS was open at a reactor coolant pump. Had DHR been lost for a sufficient period of time that boiling in the core region occurred, the vent valves would have opened to vent the steam directly to the cold legs. This would have prevented any significant reduction in pressure vessel level due to increasing pressure above the core. The location of this valve is shown in Fig. 1.

ASP Modeling Assumptions and Approach

The event has been modeled as a loss of decay heat removal during midloop with the non-running DHR train initially unavailable. Based on the 1985 loss of DHR procedure, recovery of the non-running train and the use of spent fuel pool cooling as an alternate means of providing decay heat removal are addressed as proceduralized actions. Controlled makeup to the RCS using HPI, LPI, or gravity feed from the BWST is addressed as an ad-hoc recovery action.

The event tree model is shown in Fig. 2. Based on the heatup rate specified in the LER, the time to saturation is estimated to be 83 minutes. This time period is considered more than adequate to perform the proceduralized actions which were required to open the closed DHR suction valve, DHV-39, and to implement alternate cooling using SFC system, if the valve could not be opened. Therefore, only the likelihood of equipment failure was considered when estimating branch probabilities, and not the likelihood of failing to implement required actions.

Event tree branch probabilities were estimated as follows:

1. Alternate DHR train started before saturation. A valve failure-to-open probability of 0.01/demand was used in the model. While this value is consistent with other ASP analyses, it is conservative compared to values used in NUREG-1150 efforts (3×10^{-3}/demand, see NUREG/CR-4550, Vol. 1, Rev. 1). Since both of these values include failures associated with valve operators and actuation logic, they are both probably conservative for local, manual valve operation which was actually performed during the recovery of DHR. However, since the cause of the valve failing to operate was attributed to a variety of mechanical and electrical problems, the assumption of a typical manual valve failure-to-open probability (1×10^{-4}/demand) cannot be justified.

2. Decay heat removal using the spent fuel cooling system prior to saturation. On Crystal River 3, the SFC system can be valved into the DHR system in the event that DHR pumps or heat exchangers are unavailable. This process, specified in OP-405, "Spent Fuel Cooling System," involves alignment of SFC and DHR system components to provide flow from the DH drop line, through one of the two SFC pumps and heat exchangers, and back to the RCS via the "B" DH inlet line.

   Considering the position of DHR system valves prior to the event, use of the SFC system requires the opening of two manual valves which normally isolate this system from the DHR system (SFV-89 and SFV-87), closure of two valves to isolate the spent fuel storage pools from the SFC system (SFV-8 and SFV-9), and the start of one of two SFC pumps (SFP-1A or SFP-1B). Several additional valves must be operated, but alternate series valves or parallel paths exist should these valves fail. Based on the screening probability values used in the ASP program, the probability of not initiating cooling using the SFC system is estimated to be 1.4×10^{-3}.

3. Controlled makeup to RCS using HPI or LPI or gravity feed from BWST (ad-hoc action at time of event). The use of HPI, LPI or core flood inventory to provide RPV makeup and delay the onset of core damage is not addressed in the procedures of 1986. This action has been included on the event tree as an ad-hoc action, and was assigned a failure probability of 0.1. This value is consistent with IPE requirements for non-proceduralized actions.

Analysis Results

The conditional core damge probability estimated for this event is 1.4×10^{-6}. This low value reflects the fact that an alternate, proceduralized approach for decay heat removal was available following the loss of the operating DHR train, and that the non-operating train could be recovered by local recovery of one valve.

Had this event occurred at a later time, when the current loss of DHR procedure was in effect, the conditional probability would be estimated to be below 1.0×10^{-6}. This is a result of the additional methods for decay heat removal specified in the current procedure.

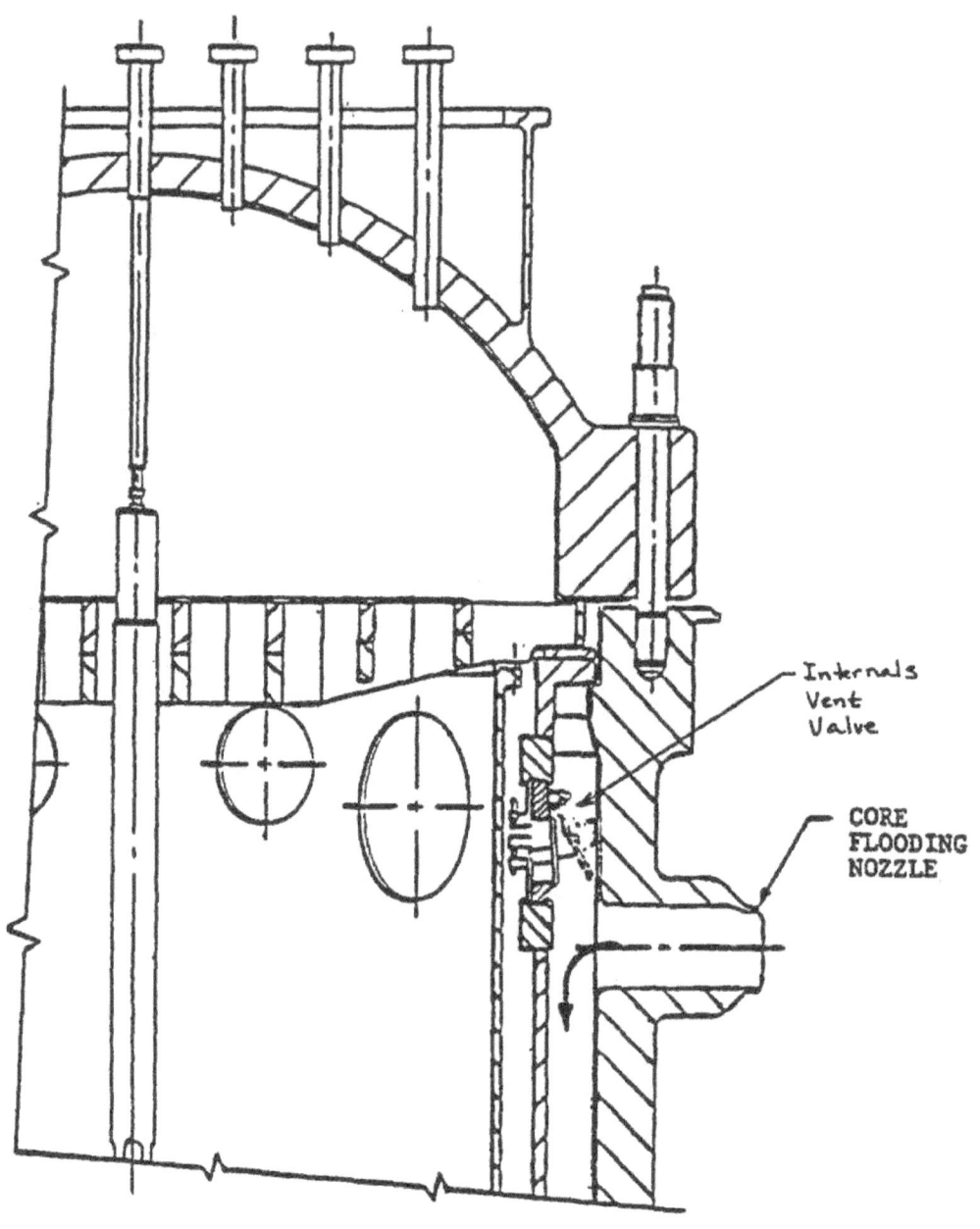

Fig. 1. Internals Vent Valve and Core Support Shield

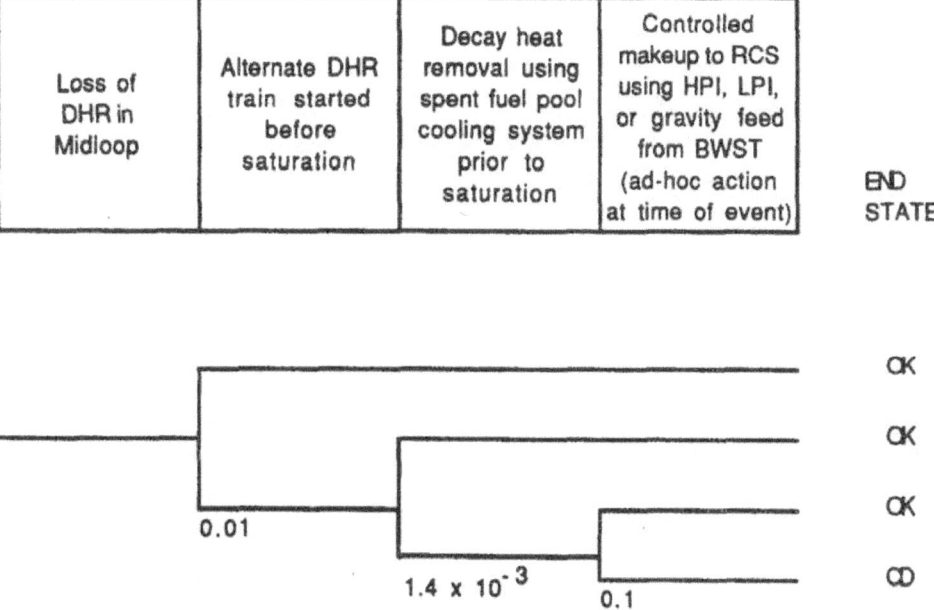

Figure 2. Event Tree Model for LER 302/86-003

ACCIDENT SEQUENCE PRECURSOR PROGRAM COLD SHUTDOWN EVENT ANALYSIS

LER No.:   323/87-005 R2
Event Description: Loss of RHR cooling results in reactor vessel bulk boiling
Date of Event: April 10, 1987
Plant: Diablo Canyon 2

Summary

During the first refueling outage, the reactor coolant system (RCS) was drained to mid-loop to facilitate the removal of the steam generator (SG) primary manways for nozzle dam installation prior to SG work. As a result of a leaking valve during a penetration leak-rate test, RCS inventory was lost. The resulting low RCS level caused vortexing and air entrainment and loss of both residual heat removal (RHR) pumps. RHR cooling was lost for ~1.5 h, during which boiling occurred. After determining that the SG manways had not been removed, the RCS was flooded by gravity feed from the refueling water storage tank (RWST) and an RHR pump restarted.

The conditional core damage probability point estimate for this event is 5.5×10^{-5}. This value is strongly influenced by assumptions concerning the operation staff's ability to implement non-proceduralized recovery actions.

Event Description

On April 10, 1987, the RCS was drained down to mid-loop to facilitate the removal of primary SG manways for nozzle dam installation prior to SG work. The plant was in the seventh day of the first refueling outage. RCS temperature was being maintained at ~87F. Local leak rate testing of containment building penetrations was also being performed.

Temporary reactor vessel water level indication was being provided by a Tygon tube manometer inside containment and two level indicators in the control room. The level alarms on the reactor water level indication system (RVRLIS) had not yet been reset to alarm at the mid-loop low level setpoint of 107'.

Reactor vessel level was being varied by draining to and feeding from the RWST via valves 8741, 8805A, or 8805B, as appropriate. Letdown was from the RHR pump discharge via valve HCV-133, and charging was by flow from the volume control tank (VCT) via the normal charging path (through a non-operating centrifugal charging pump). Once the RCS had been drained down to mid-loop (107'), level was being maintained by balancing letdown flow and makeup (charging) flow with the aid of VCT level changes. The allowed level range was from 107'0" (below which RHR pump cavitation was expected due to vortexing and air entrainment) and 108'2" (at which water could enter the channel head areas of the SGs).

RHR pump 2-2 was in service providing flow through both RHR heat exchangers (the trains were cross-tied). RHR pump 2-1 was operable but not in service. All RHR system instrumentation was in service.

Additionally:

- The safety injection (SI) pumps were electrically isolated but available for service, if manual operation of valves was performed.

- Centrifugal charging pump (CCP) 2-2 was operable and available for immediate service. CCP 2-1 and the nonsafety-related positive displacement charging pump were tagged out but were available for service.

- The RWST was available with level at approximately 97%.

- All four accumulators had been cleared and drained.

- All four SGs had a secondary side water level of approximately 73%, with the generators vented to atmosphere through the open secondary pressure relief system.

- All core exit thermocouples had been disconnected in preparation for reactor vessel head removal.

- The containment equipment hatch and personnel air lock were open. The emergency personnel hatch was closed. Various jobs were in progress inside of containment, and a continuous purge was in progress with the containment ventilation exhaust fan discharging to the plant vent.

At approximately 2010 h, a plant engineer entered containment to begin draining a containment penetration to conduct a local leak-rate test. The penetration had been previously isolated, but one of the isolation valves did not properly seat. The plant engineer did not notify the control room that he was draining the penetration. Due to the leaking isolation valve, a drain path was created between the VCT and the reactor coolant drain tank (RCDT). VCT level immediately began to decrease. The operators attempted to restore VCT level by increasing letdown flow to the VCT. This action resulted in a slow decrease in the reactor vessel water level, as indicated on the temporary RVRLIS.

Due to the apparent loss of inventory from the RCS, plant operators isolated charging and letdown flow paths at approximately 2122 h. The resulting loss of flow to the VCT caused the VCT level to decrease rapidly. The decrease in the level in RCS stopped at an indicated level on the RVLIS of 107'4".

At 2125 hours control room operators noticed that the amperage on the 2-2 RHR pump began to fluctuate. The pump was shut down, and RHR pump 2-1 was started. Amperage on the 2-1 RHR pump also fluctuated and it was shut down. Plant operators suspected vortexing or cavitation of the pumps as the cause of the pump motor amperage fluctuations. At this point both RHR pumps were stopped, RHR cooling capability was lost, and RCS heatup began. Since the core exit thermocouples had been decoupled in preparation for subsequent reactor head removal, no RCS temperature indication was available to the plant operators.

Since the apparent vortexing or cavitation of the RHR pumps was unexpected, plant operators suspected the validity of the temporary RVRLIS indication in the control room, and an operator was dispatched into the containment building to verify level indication on the Tygon tube manometer which was being used for RCS level indication inside containment.

The shift foreman, being uncertain of the status of activities involving the removal of primary side manways on the SGs, requested that the status of this work be verified. This was necessary to assure that no personnel were inside or in the vicinity of the SG channel heads or manways before he opened valves in either of two paths to allow gravity flow of water from the RWST to the RCS.

At approximately 2210 h, the control room recorder for the temporary RVRLIS began to show an increase from 107'4". (Plant operators subsequently, at approximately 2241 h, attributed the indicated increase in RVRLIS indication to steam formation in the reactor vessel head area.) Eleven min later, the control room operators received notification that the Tygon tube manometer inside containment indicated a level of between 106'9" and 107'0". At this time an attempt was made to restart RHR pump 2-1. The pump was immediately shut down due to amperage fluctuations.

At approximately 2241 h, the control room operators were notified that the SG manways had not been removed, although bolts securing some of the manways had been de-tensioned. Valves were then opened from the RWST to establish makeup to the RCS. Thirteen min later, with RCS water level indicating 111'7", plant operators successfully restarted RHR pump 2-2. Shortly following the pump start, the RHR pump discharge temperature on the control board recorder rose to approximately 220F. Within five min, the pump discharge temperature had dropped to less than 200F.

Event-Related Plant Information

RHR Design. The Diablo Canyon 2 RHR system consists of one suction pipe which draws water from one RCS hot leg, two RHR pumps, two heat exchangers, and return lines which direct cooled water back to the RCS cold legs. At Diablo Canyon, water is normally returned to all four cold legs.

RCS Level Indication and Control. When the RCS is partially drained, water level is measured by making two connections to the RCS and determining a pressure difference. The first connection is an RCS drain on the crossover pipe of Loop 4, and the second is at the top of the pressurizer. Two types of level instrumentation are used — a Tygon tube for local level indication and two differential pressure transmitters which display level in the control room on a recalibrated and relabled accumulator level instrument. The level observable in the Tygon tube was assumed to be

RCS level. The Tygon tube manometer in use during this event suffered form a number of deficiencies:

- the tube was of small diameter (which slowed response) and its installation was poorly controlled.

- the level of interest was in a high radiation area and was difficult to read.

- the Tygon tube was marked with a marking pen at approximately one-ft graduations. Water level had to be estimated by sighting structural elevation markings and transposing by eye across available cat walks, etc. to the Tygon tube.

RVRLIS level indication is influenced by RHR flow, the extent of air entrainment and temperature differentials. Level indication in the Tygon tube was further impacted by the small diameter of the tubing, which introduced significant delays in response. The utility estimated that two inches was added to indicated RVRLIS level by pumping 10% entrained air at 3000 gpm RHR flow.

RCS drain down in preparation for SG maintenance requires very close control of RCS level. Rapid draining of SG tubes requires RCS level be maintained below 107'5.5" but above 107'3.5", at which vortexing in the vicinity of the RHR suction piping connection is fully developed with an RHR flow of 3000 gpm (Westinghouse calculation). At 1500 gpm, vortexing is fully developed at 107'1.2".

Core Heatup. Bulk boiling was estimated to have occurred 45 min after loss of RHR. This was twice as fast as indicated in information available to the operators at the time of the event. Since the RCS was essentially intact, little inventory was lost, and it has been concluded (NUREG-1269, "Loss of Residual Heat Removal System") that the core would have remained covered for an extended period of time because of condensation of steam in the SGs. If the SG primary manways had been removed at the time of the event, thereby providing a vent path for the RCS, time to core uncovery is estimated to be 1.6 h after initiation of boiling, or 2.4 h total.

RHR Recovery and Supplemental RCS Makeup. Diablo Canyon procedure OP AP-16, Rev. 0, "Malfunction of the RHR System," applicable at the time of the event provided no information specifically concerning loss of RHR during mid-loop operation. General guidance was provided for loss of RHR with the reactor head in place (repressurize the RCS with the charging pumps, start a reactor coolant pump or establish natural circulation, and utilize the SGs for decay heat removal).

For this event, the RWST was full and had been used earlier to provide RCS makeup water. In addition, the SI pumps and charging pumps could be used for RCS makeup.

Analysis Approach

Core Damage Model. The core damage model considers the possibility that the loss of RPV inventory and subsequent loss of RHR could have occurred either with the RCS intact (which was the case during the event) or with the RCS vented to the containment through openings such as the SG manways.

In the event the RCS is intact, core cooling is assumed to be provided if RCS makeup is provided and if RHR is recovered or the SGs are available for steaming. For the SGs to be effective for core cooling, steam from the reactor vessel must travel to the SGs, and condensate must flow back to the vessel, as described in NUREG-1269.

If the RCS is open, then continued RCS makeup is assumed to provide core cooling success.

The event tree model is shown in Figure 1. Three core damage sequences are shown. Sequence 1 involves the situation in which the RCS is open and RCS makeup is not provided. Sequences 2 and 3 involve a closed RCS. In sequence 2, RCS makeup is provided, but both RHR recovery fails and the SGs are unavailable for core cooling. In sequence 3, RCS makeup fails.

Branch probabilities were estimated as follows:

a. RCS Open. At the time of the loss of RHR, the RCS was closed. However, the SG manways were scheduled to be removed at about the time of the event. The likelihood of the RCS being open was assumed to be 0.5.

b. RCS Makeup. The likelihood of failing to maintain RCS makeup for decay heat removal if the RCS was open was estimated based on crew error probabilities developed from time reliability correlations and shown in Figure 2. Four types of crew response are addressed: (1) response based on detailed operating procedures, (2) trained knowledge-based performance, (3) typical knowledge-based performance, and (4) knowledge-based performance during very unusual events. Figure 2 was developed from curves appropriate to in-control room action, and the response time was skewed 20 min to account for recovery outside the control room. Typical knowledge-based response was assumed for the event (the operating procedure provided no information concerning mid-loop operation). For the estimated 2.4 h to core uncovery, a crew error probability of 1.0×10^{-4} is indicated.

For cases in which the RCS was closed, restoration of RCS level to allow RHR pump restart was considered to be a part of normal RHR recovery actions. The failure probability for equipment associated with restoration of RCS level was estimated to be 1.0×10^{-5}.

c. RHR recovery. Recovery of RHR was effected by starting RHR pump 2-2 after RCS level was recovered. It was assumed that RHR pump 2-1 could also have been used, although venting might have been required. Failure of RHR would therefore require failure of both RHR pumps to start and run. Based on probability values typically used in the ASP program, a branch probability of 3.4×10^{-4} is estimated.

d. SGs provide core cooling. During this event, SG inventories were at ~73%. Since the secondary relief system was open, continued decay heat removal could be provided as long as SG makeup was available. For this analysis, it was assumed that the motor-driven AFW pumps were available for SG injection. (SG makeup would only have been required after a considerable period of time, considering the water level in the SGs at the start of the event.) A branch probability of 3.4×10^{-4} was utilized in this analysis.

Analysis Results

The estimated core damage probability associated with the loss of RHR cooling at Diablo Canyon is 5.5×10^{-5}. This value is strongly influenced by assumptions concerning operator action during the event.

Substantial uncertainty is also associated with this estimate. Provided the RCS was intact and the SGs were available for decay heat removal, an extended period of time was available to effect recovery. If the RCS was open, 2.4 h were still available for recovery. However, recovery actions were not proceduralized at the time of the event.

The impact of different assumptions concerning the time after shutdown, the status of the RCS, and ability to cool the core using SGs as described in NUREG-1269 are shown below.

Assumption	Revised Core Damage Probability
Event occurs two days after shutdown (time to boil estimated to be 0.13 h, time to core uncovery with open RCS estimated to be 1.0 h.).	1.3×10^{-3}
SG manways removed.	1.0×10^{-4}
Natural circulation cooling using SG ineffective.	1.8×10^{-4}

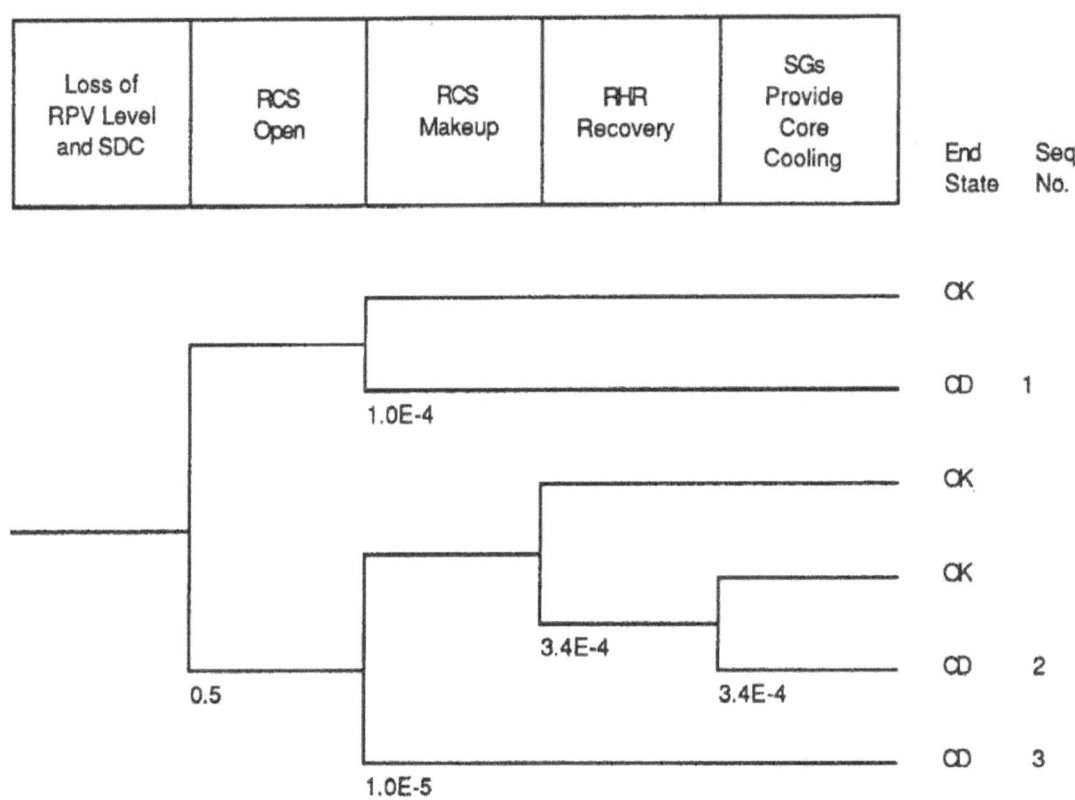

Fig.1. Event tree model for LER 323/87-005 R2

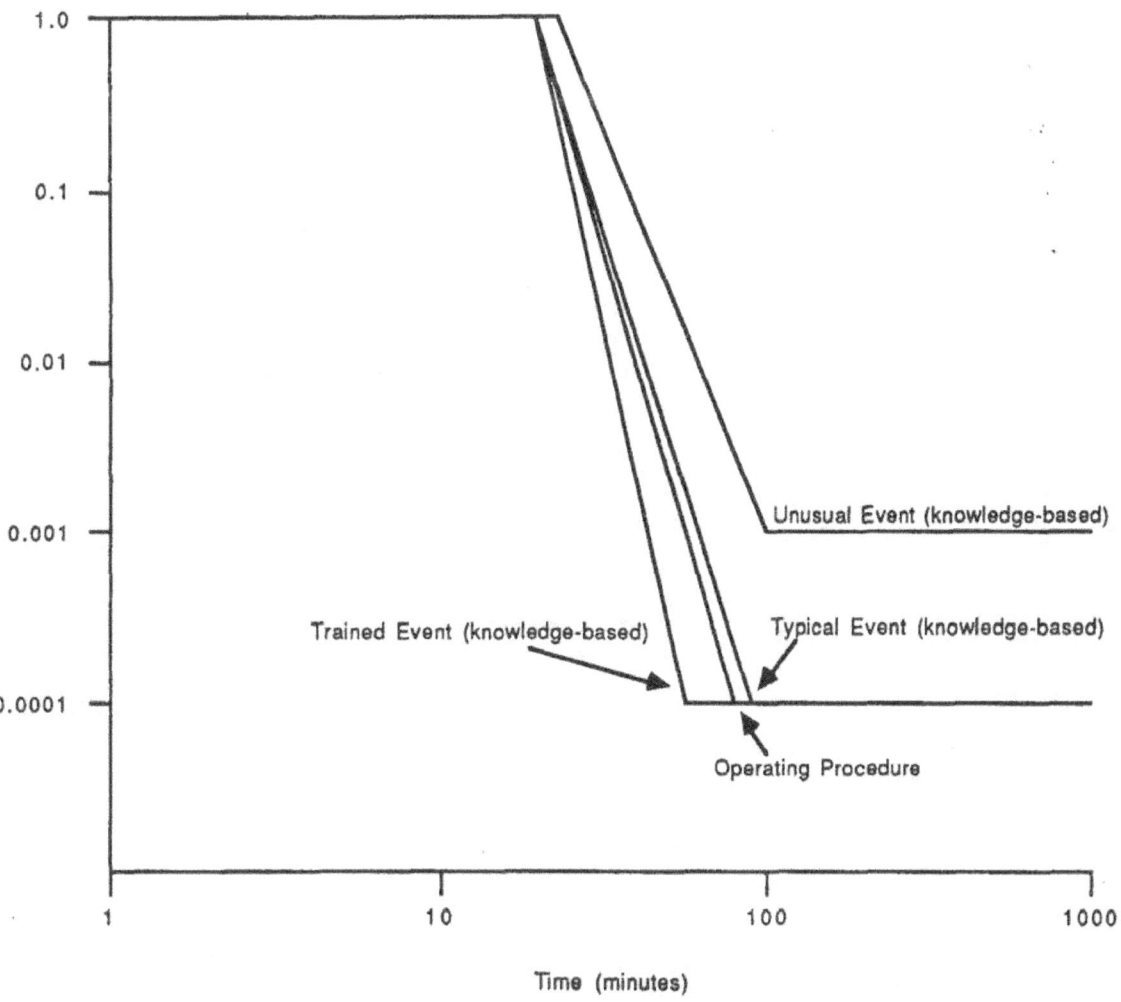

Fig. 2. Probability of not implementing RCS makeup

ACCIDENT SEQUENCE PRECURSOR PROGRAM COLD SHUTDOWN EVENT ANALYSIS

LER No: 382/86-015
Event Description: Localized boiling during mid-loop operation
Date of Event: July 14, 1986
Plant: Waterford Unit 3

Summary

While draining the reactor coolant system (RCS) to mid-loop in preparation for replacement of a RCS pump seal, RCS level dropped below mid-loop and the operating shutdown cooling (SDC) pump [low-pressure safety injection (LPSI) pump "B"] cavitated. Approximately 4 h were required to restore SDC (level was restored approximately 40 min after recognition that the "B" LPSI pump was cavitating). During this period, local boiling was occurring in the reactor vessel.

The conditional core damage probability estimate for this event is 2.1 x 10^{-4}. This value is strongly influenced by assumptions concerning the operation staff's ability to restore SDC using non-proceduralized pump jogging and the availability of the steam generators (SGs) as an alternate means of removing decay heat.

Event Description

On July 14, 1986, at 0113, personnel drained the RCS to mid-loop (13'4" elevation at centerline of hot-legs) in preparation for replacement of the seal package for the "2A" reactor coolant pump. The water was being drained into the refueling water storage pool (RWSP) via

(1)   the LPSI pump "B" mini-recirculation valves SI-120B and SI-121B (this was not specified by procedure), and

(2)   the holdup tanks via the chemical and volume control system (CVCS) purification exchangers through valve SI-423.

Personnel secured draining the RCS (incorrectly) at 0113 by just closing SI-423. Operations personnel neglected to close SI-120B and SI-121B; this resulted in RCS inventory being pumped into the RWSP.

A temporary Tygon tubing line was being used to measure RCS level. Throughout the draining operation, personnel experienced problems with the Tygon tubing. Positive pressure in the RCS was maintained by a nitrogen blanket. However, nitrogen could not be added fast enough to compensate for the drain down. Therefore, a slight vacuum existed in the RCS. This slight vacuum caused indicated RCS level to fluctuate. Because of this, operators did not trust the level indication.

To obtain an accurate reactor vessel level indication, operations personnel began venting the RCS. The process was complicated by the need to substitute local operators because the original operator was suffering from heat prostration. Upon completion of the venting process, the indicated vessel level fell to 9 ft (well below the hot leg). As a precaution, operations personnel initiated charging flow. Since the LPSI pump "B" was operating satisfactorily and the reactor vessel monitoring system indicated a higher level, operations personnel felt that the local indication was inaccurate.

At 0317, LPSI pump "B" began to cavitate. The pump was immediately secured thus terminating shutdown cooling flow. At this time, personnel realized they neglected to close valves SI-120B and SI-121B and immediately closed the valves. In order to fill the RCS with LPSI pump "A", valve SI-109A was opened. LPSI train "A" was originally aligned for SDC; however, by opening SI-109A LPSI train "A" was re-aligned to inject water into the RCS from the RWSP. The RCS was being refilled at approximately 600 gpm. At 0351, vessel level was observed to be just below centerline of the hot leg.

At 0400, conditions within the RCS indicated that local boiling was occurring (i.e., core exit thermocouples were reading 223F). Several attempts were made to start LPSI pump "B"; however, cavitation persisted (probably due to air and/or steam binding). [Note: NRC Inspection Report 50-382/86-15 notes that LPSI pump "A" also cavitated when it was started.]

Operations personnel attempted to restore SDC by jogging the "A" and "B" LPSI pumps while cycling their respective warm up valves, SI-135A and SI-135B. Therefore, intermittent flow was being established by jogging the pumps. By opening SI-135A and SI-135B when jogging the pumps, some of the water was being diverted back to the LPSI pump suction, thus priming the pumps. This operation continued until approximately 06:58 when LPSI pump "A" was secured and SDC was re-established with the "B" LPSI pump.

Fig. 1 contains a simplified drawing of the RHR system.

Event-Related Plant Information

The Loss of Shutdown Cooling procedure applicable at the time of the event (OP-901-046 Rev. 2) addressed both system leakage and loss of SDC flow, but provided little detailed guidance.

If RCS level indications were not stable (decreasing), the procedure specified that LPSI flow was to be initiated. If LPSI flow could not maintain RCS level, then HPSI was to be initiated. If HPSI had been used to recover RCS level and that level had returned to normal, then the steam generators (SGs) were to be used for decay heat removal provided the RCS was pressurized. If the RCS was depressurized (presumably open), then containment cooling was to be maximized. If the LPSI pumps were used for RCS makeup, then one pump was to be stopped and the suction of the other shifted to partially take suction on the RCS via the RCS drop line.

For a loss of SDC, the procedure required use of the SGs for decay heat removal if no RHR pumps could be returned to service. If loss of flow was due to air binding, the procedure required the shutdown priming system be placed in service.

The LPSI pumps serve two functions. One of these is to inject large quantities of borated water into the RCS in the event of a large pipe rupture. The other function of the LPSI pumps is to provide shutdown cooling flow through the reactor core and shutdown cooling heat exchanger for normal plant shutdown cooling operation or as required for long-term core cooling for small breaks. During normal operation the LPSI pumps are isolated from the RCS by motor-operated valves. When performing their safety injection function, the pumps deliver water from the RWSP to the RCS, via the safety injection nozzles. Sizing of the LPSI pump is governed by the shutdown cooling function.

The high-pressure safety injection (HPSI) pumps primary function is to inject borated water into the RCS if a break occurs in the RCS boundary. The HPSI pumps are also used during the recirculation mode to maintain borated water cover over the core for extended periods of time. For long term core cooling, the HPSI pumps are manually realigned from the main control room for simultaneous hot and cold leg injection. This insures flushing and ultimate subcooling of the core independent of break location.

The HPSI and LPSI pumps are located in rooms in the lowest level of the reactor auxiliary building. This location maximizes the available net positive suction head (NPSH) for the safety injection pumps.

During the July 14, 1986 event, one LPSI pump was used to restore RCS level (This is required by the RCS leakage portion of the procedure, but not by the loss of SDC portion. Erratic SDC flow is an indication for the RCS leakage portion of the procedures). However, the vacuum priming system was apparently not used to vent the LPSI pump suction piping even though required by the loss of SDC portion of the procedure. Instead, flow through the LPSI pump warm-up lines was used, together with jogging the pumps, the re-establish shutdown cooling flow. This process took three hours. (The difficulty with this can be seen from the RCS elevation shown in Fig. 2. The LPSI pump suction piping raises in a U-bend 9 ft above the bottom of the hot leg. Once this U-bend is voided, it could not be easily refilled without the use of the vacuum priming system. However, during this event, hot leg temperatures were greater than 212F, and the vacuum priming system could not have been used to evacuate the loop seal.)

In addition to the LPSI and HPSI pumps specified by procedure, the containment spray system (CSS), safety injection tanks (SITs), and charging pumps could be used to inject borated water into the RCS on an ad-hoc basis. A brief description of these systems follows.

The CSS consists of two independent and redundant loops each containing a spray pump, shutdown heat exchanger, piping, valves, spray headers and spray nozzles. The system has an injection mode and a recirculation mode. Containment spray pumps can be aligned to inject into the same cold-leg RCS piping as LPSI and HPSI.

Four SITs are used to flood the core with borated water following depressurization as a result of a loss-of-coolant accident (LOCA). Each SIT has a total volume of 2,250 ft^3 and a water volume of from 1,679 ft^3 to 1,807 ft^3 (12,600 gal to 13,517 gal) of borated water at a pressure of 600 psig (235 to 300 psig in shutdown). Each SIT is piped into a cold leg of the RCS via a safety injection nozzle located on the RCS piping near the reactor vessel inlet. Although the SIT isolation valves are closed when RCS pressure is down to 377 psig the operator can open these valves.

A method available for injection of unborated water immediately is one of three positive displacement charging pumps (capable of injection at approximately 44 gpm each). The other two charging pumps could be "racked" in and started in a short period of time.

The three positive-displacement charging pumps (44 gpm each) can also be used for RCS injection. During cold shutdown, two of these pumps are normally depowered, but could be restored to power by racking in the pump breakers.

Analysis Approach

The event tree model developed for this event is shown in Fig. 3. This model is based on the procedure in effect at the time of the event and includes the use of both HPSI and LPSI for RCS makeup. If the RCS is open to containment, then continued makeup provides core cooling success. If the RCS is closed (as it was during this event), then recovery of SDC or use of the SGs (either by steaming or through a bleed and feed operation involving the blowdown system) is also required for core cooling success.

Branch probabilities were estimated as follows:

a. RCS open. During this event, the RCS was closed. A branch probability of 1.0 was utilized.

b. RCS makeup. Success of either LPSI or HPSI will provide adequate makeup to the RCS.

In this event, one LPSI pump had been secured because it was cavitating. The branch probability for failure of LPSI was developed under the assumption that only one LPSI pump was considered to be available. For LPSI success, that pump must start and run and its associated RWSP isolation valve must open. The failure probability for LPSI makeup is estimated to be 6.8×10^{-3}, using component failure probabilities typical of other calculations in the ASP program.

Three HPSI pumps are normally available but depowered while in cold shutdown. These pumps provide flow to the four RCS cold legs through parallel, normally closed, motor-operated injection valves (two per cold leg). For HPSI success, one pump must start and run, and one associated injection valve must open. Based on the probabilities employed in the ASP program, the failure probability for HPSI injection is estimated to be 1.5×10^{-4}.

Combining these values results in an overall failure probability for RCS makeup of 1.0×10^{-6}.

c. RHR recovery. Recovery of RHR required three hours and involved use of the LPSI pump warmup lines in conjunction with LPSI pump jogging, which was inconsistent with the procedure. A failure probability of 0.3 was assumed in the analysis.

d. SGs provide core cooling. During this event, both SGs were available for heat removal. Emergency feedwater (motor-driven pumps) and the atmospheric dump valves were available. Based on probability values employed in the ASP program, a failure probability of 6.8×10^{-4} is estimated.

Analysis Results

The estimated conditional core damage probability associated with the loss of RCS level and RHR cooling at Waterford is 2.1×10^{-4}. This value is strongly influenced by the assumption that recovery of RHR cooling by repeated LPSI pump jogging, as was done during the event, was marginal. The dominant sequence involves failure to recovery RHR and failure to remove decay heat using the SGs.

The event conditional probability is also strong influenced by the fact that the SGs were available for decay heat removal. If this were not the case — for example, if the event had occurred during an extended outage when extensive work was being performed on the secondary side — a significantly higher core damage probability would be estimated.

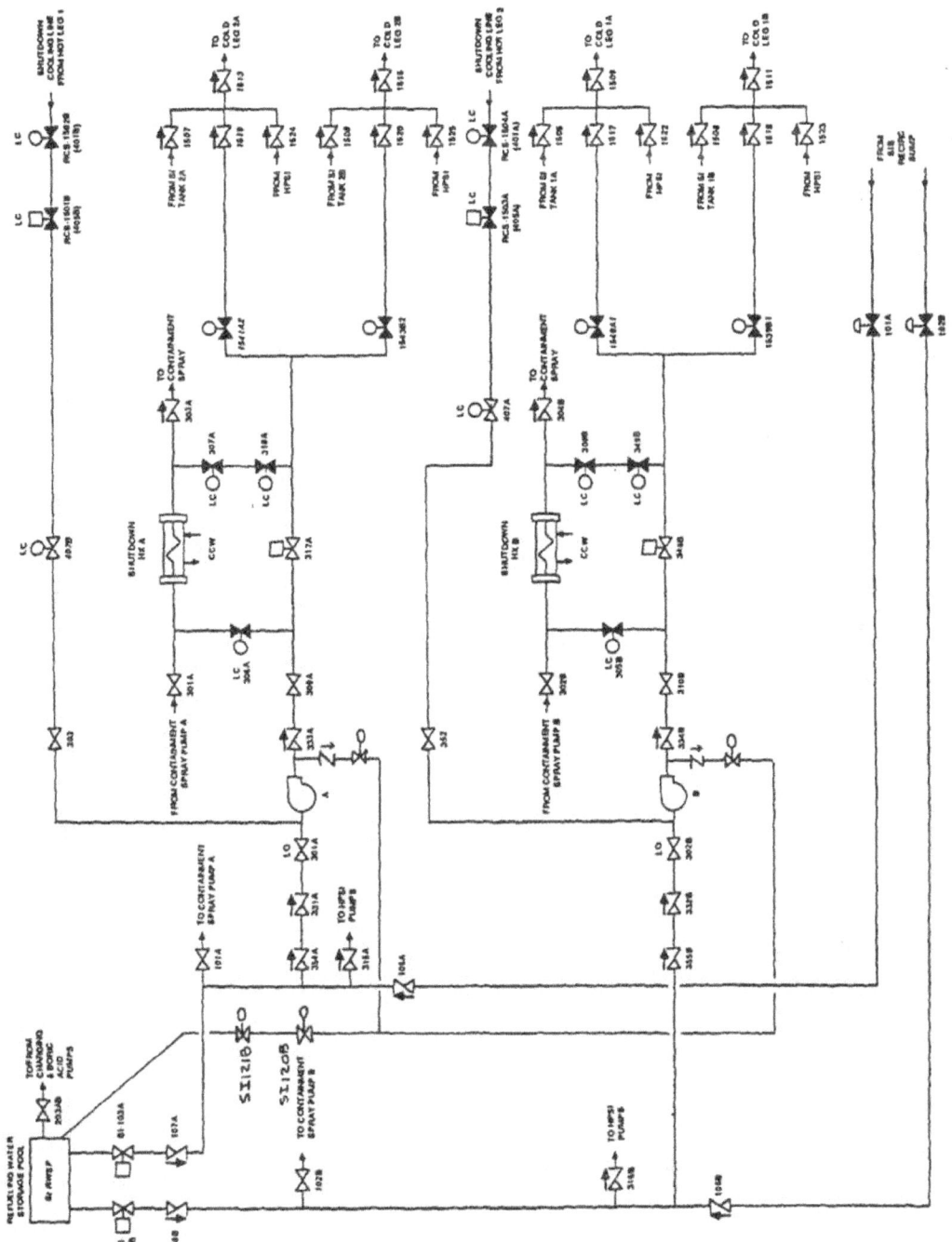

Fig. 1. Simplified drawing of the Waterford RHR system

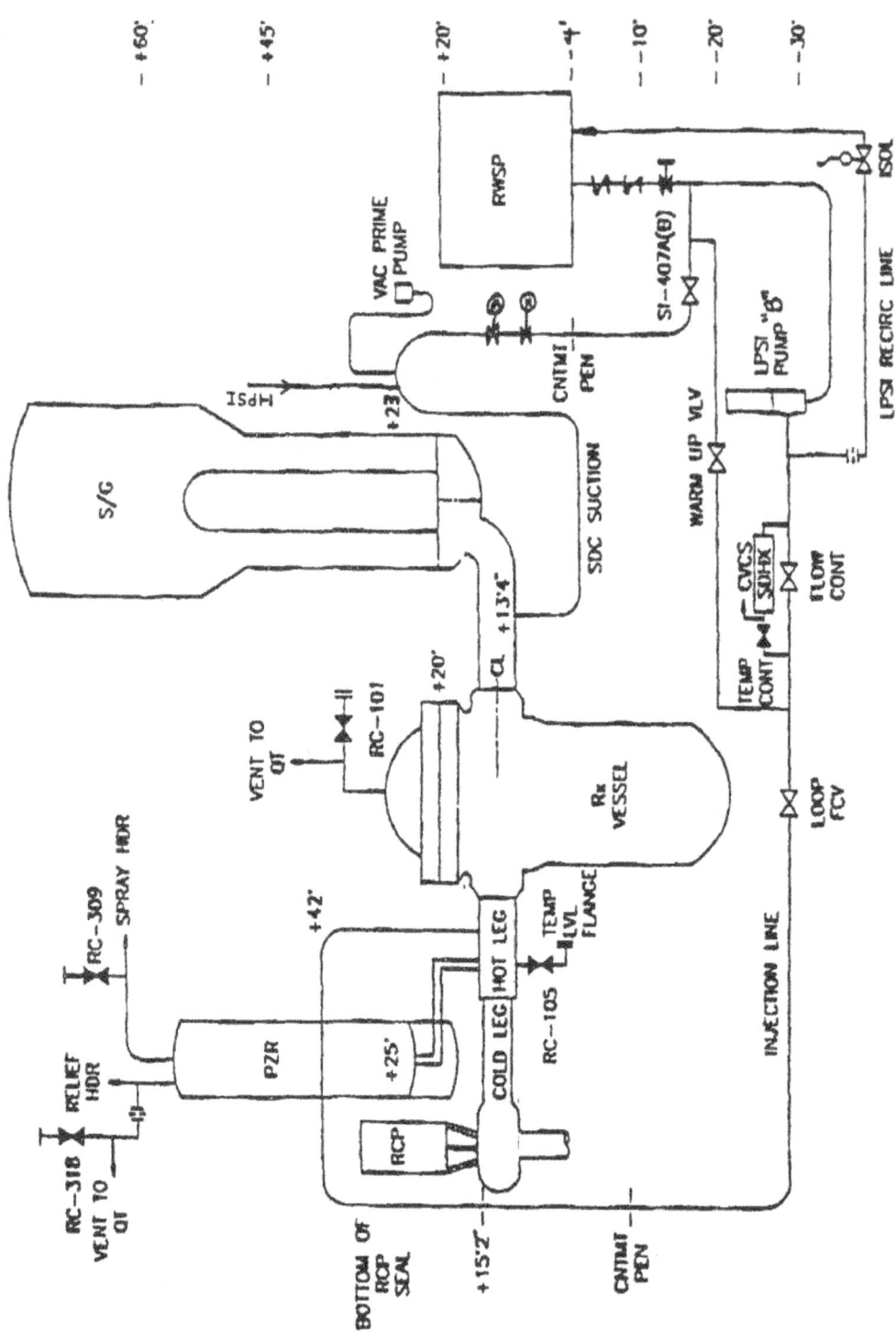

Fig. 2. RCS/SDC position and elevation reference

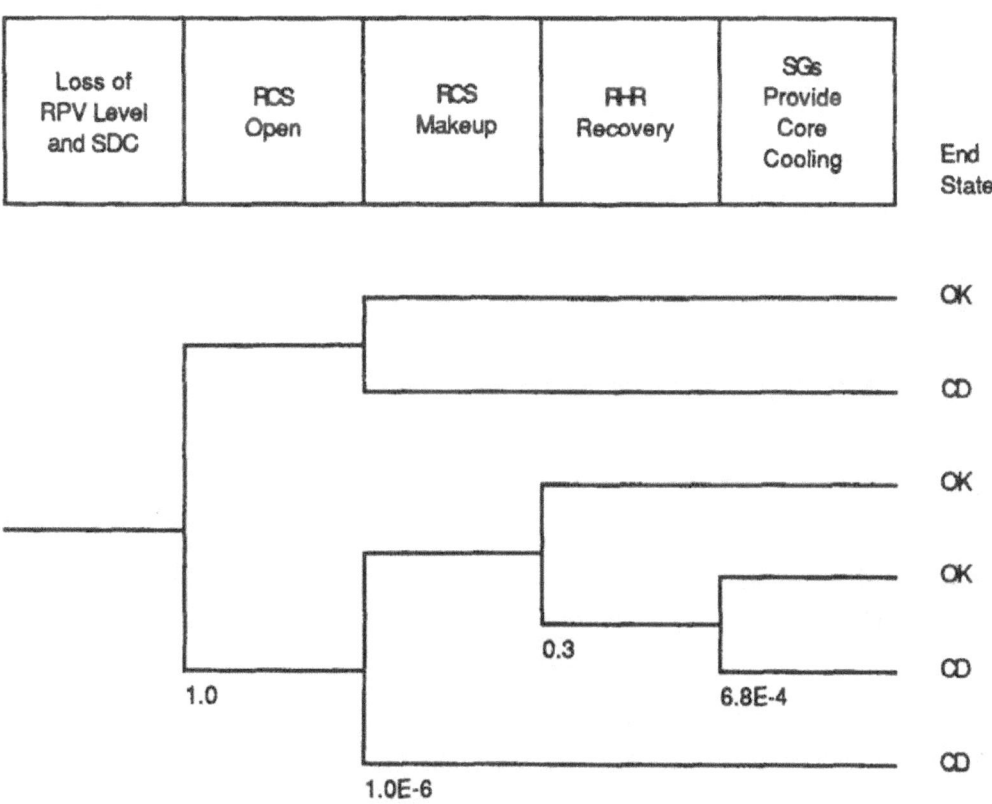

Fig. 3. Event Tree Model for LER 382/86-015

ACCIDENT SEQUENCE PRECURSOR PROGRAM COLD SHUTDOWN EVENT ANALYSIS

LER No.:   387/90-005
Event Description: RPS bus fault results in loss of normal shutdown decay heat removal
Date of Event: February 3, 1990
Plant: Susquehanna 1

Summary

On February 3, 1990, a loss of reactor protection system (RPS) bus B occurred at Susquehanna 1 during RPS bus breaker testing, a result of a short to ground in a DC distribution panel. The loss of the RPS bus prevented recovery of residual heat removal (RHR), which had been previously isolated for the breaker testing, for over five h. The conditional probability of subsequent severe core damage estimated for the event is 2.7×10^{-5}. Dominant sequences are associated with failure to implement alternate core cooling strategies in the event that RHR could not be recovered in the short term. The calculated probability is strongly influenced by estimates of the likelihood of failing to recover initially faulted systems over time periods of 6-24 h. These estimates involve substantial uncertainty, and hence the overall core damage probability estimated for the event also involves substantial uncertainty.

Event Description

On February 1, 1990, Susquehanna 1 was shutdown due to a leak in the main turbine hydraulic control system. The leak was repaired and preparations for startup began. The plant was in operational condition 4 (shutdown with reactor coolant temperature less than 200F) with the "A" loop of the RHR system in service in the shutdown cooling (SDC) mode.

At 1555 on February 3, 1990, with reactor coolant temperature at approximately 125F, the RHR system was removed from service as part of preparations for performing a semi-annual functional test of the RPS electrical protection assembly (EPA) breakers. The EPA breakers, two in series for each RPS bus source, ensure that the power supplied is within the voltage and frequency design specifications of the RPS by automatically tripping open when a power source is outside of this specification. The normal power supply to each of the RPS buses (A and B) is a dedicated motor generator set and the alternate is a dedicated voltage regulating transformer. RHR is taken out of service during this surveillance because isolation signals to the RHR SDC suction valves, HV-151F008 and 9, are initiated when the RPS distribution buses are de-energized during the test. With the exception of the EPA breaker functional test, all surveillances required for startup were complete.

The EPA breaker functional test was in progress. All EPA breakers had been demonstrated to be functioning properly and only restoration activities remained to be performed. The last two EPA breakers (normal supply to RPS bus "B") had been tripped open satisfactorily. All other EPA breakers had been reset and closed previously in the test.

At 1725 on February 3, 1990, with reactor coolant temperature at 188F, attempts to restore normal power to RPS bus "B" by resetting and closing the last two EPA breakers tested were unsuccessful. When attempts were made to transfer RPS bus "B" to its alternate supply, the alternate supply EPA breakers also tripped open. A consequence of not being able to restore power to RPS bus "B" is the inability to restore RHR SDC due to the fact that the isolation signals to the reactor vessel suction valves, which are common to both loops of RHR, were still present.

The loss of RPS bus "B" was caused by a short circuit to ground in the RPS bus "B" distribution panel. This occurred when a copper mounting bolt (also used as a conductor) for one of the bus output breakers shorted to the breaker mounting baseplate. The cause of the fault was a combination of the breaker mounting/termination configuration design and the fact that the length of the insulating sleeve, as supplied by the vendor, was insufficient to completely insulate the mounting/conductor bolt from the baseplate.

The plant implemented the existing loss of shutdown cooling procedure, ON-149-001.

The sequence of events following the loss of the RPS bus was as follows:

Time	Event
1753	Reactor coolant temperature exceeded 200F, which resulted in entry into operational condition 3 (hot shutdown). ALERT declared.
1840	The "B" loop of RHR was placed in service in the suppression pool cooling mode in preparation for manually opening SRVs, as required by procedure ON-149-001. The suppression pool temperature was 63F.
1846	With the reactor coolant at 230F and reactor vessel pressure at 10 psig, the "A" safety relief valve (SRV) was opened.
1923	With the reactor coolant at 245F and reactor vessel pressure at 15 psig, the "B" SRV was opened.
1925	The RPS EPA breakers were reset and power was restored to RPS bus "B" following repairs of the short circuit to ground in the RPS bus "B" distribution panel.
1947	With the reactor coolant at 250F and reactor vessel pressure at 19 psig, the "C" SRV was opened which stabilized reactor coolant temperature at 253F.

2240	The reactor water cleanup system, which had also received isolation signals when RPS bus "B" was de-energized during the EPA breaker test, was returned to service.
2302	The "A" loop of RHR was placed in service in the shutdown cooling mode.
2322	With the reactor coolant at 233F and reactor vessel pressure at 12 psig, the "C" SRV was closed.
2324	The "B" SRV was closed.
2327	The "A" SRV was closed.
0015-0024 (Feb. 4, 1990)	With reactor coolant at 192F, the unit was declared to be in operational condition 4 (cold shutdown), the operating recirculation pump was secured, and the ALERT was terminated.
0200 (Feb. 4, 1990)	The "B" loop of RHR, which was providing suppression pool cooling, was taken out of service. Maximum suppression pool temperature during the event was 69F.

During the event, reactor vessel water level was maintained at greater than 87" [248" above top of active fuel (TAF)] using the control rod drive (CRD) system as the source of water makeup.

Following the event, Pennsylvania Power & Light removed the existing GE type TEB-111100 circuit breakers and associated mounting plate in the RPS distribution panels on both Susquehanna units and replaced them with GE 277V distribution panels and GE type TEY-1100 circuit breakers. In addition, an investigation was conducted to determine if other similar breaker mounting configurations existed in the plant, and it was concluded that there were none. The utility stated that this investigation involved document searches, panel walkdowns, personnel surveys, and vendor assistance.

ASP Modeling Approach and Assumptions

Event Tree for Loss of RHR

An event tree model of sequences to core damage given a total loss of boiling water reactor (BWR) shutdown cooling was developed based on procedures and outage planning information developed by Pennsylvania Power & Light Company (Procedure ON-149-001, Loss of RHR Shutdown Cooling Mode, September 7, 1990, and NSAG Project report 4-90, Outage Planning Information, October 17, 1990). While the references are specific to Susquehanna, the resulting event sequences are considered applicable to most contemporary BWRs.

The event tree is shown in Fig. 1. The following comments are applicable to this event tree:

a. Core damage end state. Core damage is defined for the purpose of this model as reduction in reactor pressure vessel (RPV) level above the TAF or failure to remove heat from the suppression pool in the long term. With respect to RPV inventory, this definition may be conservative, since steam cooling may limit clad temperature increase in some situations. However, choice of TAF as the damage criterion allows the use of simplified calculations to estimate the time to an unacceptable end state.

b. Short-term recovery of RHR. All historic losses of RHR have been recovered before RPV level would have dropped to below TAF. Including RHR recovery allows operational events to be more realistically mapped onto the event tree model. Short-term RHR recovery can be delayed if a recirculation pump can be started or if RPV level can be raised to permit natural circulation. Availability of RPV injection to raise water level for natural circulation is included in the model.

c. Successful termination of the loss of RHR is defined as recovery of RHR or provision of alternate decay heat removal via the suppression pool or main condenser, or, if the head is removed, via refueling cavity boiling. Short-term decay heat removal methods (such as feed with bleed to a tank) with subsequent long-term recovery of RHR, is not addressed in the event tree, although such an approach can provide additional time to implement a long-term core cooling approach.

d. Three pressure vessel head states are addressed in the event tree: head on and tensioned, head on and detensioned, and head off. If the head is on and tensioned, then decay heat removal methods which require pressurization are assumed to be viable. If the head is on, but detensioned, then failure to maintain the RPV depressurized is also assumed to proceed to core damage (this assumption is conservative). If the head is off, then makeup at a rate equal to boil-off is assumed to provide core cooling.

e. Four makeup sources are shown on the event tree: LPCI, core spray, CRD flow and the condensate system. Branches for these sources are shown before short-term RHR recovery. This is because injection from any source to raise RPV level and allow natural circulation substantially increases the amount of time available for recovery of RHR. The four makeup sources have been placed before RHR recovery to address this issue, even though the need for significant flow from these systems is only required if RHR is not recovered (the event tree has been structured to correctly address the need for makeup if RHR is not recovered).

It should be noted that the loss of shutdown cooling procedure and the outage planning document identify other makeup and heat removal methods which have not been included on the event tree. Some of these would not have been effective at the decay heat levels which existed during the event. Others are short-term measures which eventually require transfer of decay heat to the ultimate heat sink. Additional sources of injection have not been modeled since loss of injection sequences are already of very low probability (see Fig. 2).

f. Short-term recovery of RHR is assumed to successfully terminate the loss of RHR. In the event that RHR cannot be recovered, then alternate core cooling sequences are included in the event tree. If the head is tensioned, these involve allowing the RPV to repressurize, opening of at least one SRV, and dumping decay heat to the suppression pool. If the condenser and condensate system are available, then decay heat can also be dumped to the condenser. If the head is detensioned, then decay heat must be removed without the RPV being pressurized. This requires opening of at least three SRVs and recirculating water to the suppression pool using the core spray or low-pressure coolant injection (LPCI) pumps. For all cooling modes involving the suppression pool, suppression pool cooling must be initiated in sufficient time to prevent the suppression pool from exceeding its temperature limit. If the head is removed, then any makeup source greater than ~200 gpm, combined with boiling in the RPV, will provide adequate core cooling.

Figure 1 includes the following core damage sequences:

Sequence	Description

Sequences with the Head Tensioned

103	Unavailability of long-term heat removal from the suppression pool with failure to recover RHR but following successful alternate short-term decay heat removal using LPCI or core spray injection and relief to the suppression pool via one or more SRVs.
104	Failure to recover RHR and failure to initiate alternate short-term decay heat removal due to unavailability of the SRVs for relief to the suppression pool.
107	Similar to sequence 103 except LPCI and core spray are unavailable. RPV injection provided using CRD flow.
108	Similar to sequence 104 except LPCI and core spray are unavailable. RPV injection provided using CRD flow.
112	Unavailability of long-term heat removal from the suppression pool with failure to recover RHR but following successful alternate short-term decay heat removal using the condensate system for injection and relief to the suppression pool via one or more SRVs. Relief to the suppression pool is required in this sequence because the main condenser is unavailable as a decay heat removal mechanism.
113	Failure to recover RHR and failure to initiate alternate short-term decay heat removal due to unavailability of the SRVs for relief to the suppression pool and unavailability of the main condenser as a decay heat removal mechanism.

| 115 | Failure to recover RHR and unavailability of LPCI, core spray, CRD flow and the condensate system to raise RPV level to provide for natural circulation. The time available to recover RHR in this sequence is less than for sequences with RPV injection unless a recirculation pump can be started, since RPV level cannot be raised to provide for natural circulation cooling. |

Sequences with the Head Detensioned

118	Unavailability of long-term heat removal from the suppression pool with failure to recover RHR but with successful alternate decay heat removal using LPCI or core spray injection with discharge to the suppression pool using three or more SRVs.
119	Failure to recover RHR and failure to initiate alternate short-term decay heat removal due to unavailability of three or more SRVs for relief to the suppression pool.
121	Failure to recover RHR with unavailability of LPCI and core spray for alternate decay heat removal. CRD flow provides sufficient water to raise RPV level and allow natural circulation, extending the time available to recover RHR.
123	Similar to sequence 121 except CRD flow is also unavailable. Condensate is used to increase RPV level and allow natural circulation.
125	Failure to recover RHR without RPV injection to extend RHR recovery time.

Sequence with the Head Removed

| 129 | Unavailability of LPCI, Core Spray, CRD flow and condensate for RPV makeup. Core damage in the long term if a supplemental makeup source cannot be provided. |

Branch Probabilities

Head Status. For the operational event in question, the head was on and tensioned. A review of BWR refueling outages over the last five years indicates a distribution of outage durations with peaks at 66 and 104 d. These values represent a mix of 12 mth and 18 mth refueling cycles. Assuming (1) the lower peak is more representative of a yearly refueling outage duration (and that the mean length of a yearly outage is relatively close to the peak), (2) that the fraction of time with the head on is about the same as with the head off, (3) that two d of the outage are not at cold shutdown, and (4) that the total time during an outage that the head is on but detensioned is approximately two days, results in the following time periods for the three head states over a period: head on, 31 d; head detensioned but on, 2 d; and head off, 31 d.

In addition to refueling outages, there are typically three outages of an average length of 5.6 d. If we again assume two days per outage not at cold shutdown, and assume that during the remainder of the time the plant is at cold shutdown with the head on, the following overall fractions of time for the three head states are estimated:

head on	0.56
head on but detensioned	0.03
head off	0.41

LPCI or CS Flow Available. To simplify the estimation of the probability of failure of suppression pool cooling (which is dependant on the status of LPCI), only the probability of failure of core spray was used to estimate this branch probability. For Susquehanna, the core spray system consists of two trains. Each train includes two parallel pumps with a single, normally open motor-operated suction valve and a single normally-closed discharge (RPV injection) valve. The pump suction source is normally the suppression pool. Assuming that normally-open valves and check valves do not contribute substantially to system unavailability, the equation for failure of core spray is therefore

$$(CS\text{-}P1A*CS\text{-}P1C+CS\text{-}5A)*(CS\text{-}P1B*CS\text{-}P1D+CS\text{-}5B).$$

Reducing this equation results in the following minimal cutsets:

CS-P1A	CS-P1B	CS-P1C	PS-P1D
CS-P1A	CS-P1C	CS-5B	
CS-P1B	CS-P1D	CS-5A	
CS-5A	CS-5B		

Applying screening probabilities of 0.01 for failure of a motor-driven pump to start and run and failure of a motor-operated valve to open; 0.1, 0.3 and 0.5 for the conditional probabilities of the second, third and fourth similar components to operate, and a likelihood of 0.34 of not recovering a failed core spray system in the short-term results in an overall system failure probability estimate of 4.0×10^{-4}.

If only one train is available, as would be the case of one division was out-of-service for maintenance, the core spray system failure probability (using the same approach as above) is estimated to be 3.7×10^{-3}.

CRD Flow Available. At cold shutdown pressures, one of two CRD pumps can provide makeup. Since one pump is typically running, the system will fail if that pump fails to run or if the other (standby) pump fails to start and run. Assuming a probability of 0.01 for failure of the standby CRD pump to start, and 3.0×10^{-5}/hr for failure of a pump to run, results in an estimated failure probability for CRD flow of 2.5×10^{-6}. In this estimate, a short-term non-recovery likelihood of 0.34 was applied to the non-running pump failure-to-start probability, consistent with the approach used to estimate the failure probability for the core spray system. A mission time of 24 h was also assumed.

If only one train is available (because of maintenance on the opposite division), then the CRD failure probability is estimated to be 7.2×10^{-4}.

Condensate Available. While the condensate pumps can provide more than adequate makeup, they are often unavailable during a refueling outage because of work on the secondary system. For this analysis, it was assumed that the condensate system is unavailable during a refueling outage once the plant enters cold shutdown. During a non-refueling outage, the probability of the condensate system being unavailable was assumed to be 0.1. This results in an overall unavailability, based on the fraction of cold shutdown events which are refueling-related (see Head Status), of 0.87. Since the event at Susquehanna did not involve a refueling outage, an unavailability of 0.1 was assumed.

RHR (SDC) Recovered (Short-Term). For Susquehanna, RHR can be restored to service provided RPV level is greater than the low-level isolation level and RPV pressure is less than the high pressure isolation pressure, and, of course, the cause of the initial loss of RHR is repaired.

For event tree branches with the head on and for which reactor vessel (RV) inventory was increased to provide for natural circulation, RHR must be recovered prior to RV pressure reaching the high pressure isolation setpoint (98 psig at Susquehanna), which would prevent opening the suction line isolation valves and restoring RHR. Once the high-pressure isolation setpoint is reached, operation of at least one SRV is assumed to be required, and the sequence proceeds with RPV depressurization and the use of RHR in the suppression pool cooling mode to remove decay heat. In estimating the probability of not recovering RHR (SDC), the time period of concern for these sequences is from initial loss of RHR until the high-pressure isolation setpoint is reached. (Approximately 7.5 h from the loss of RHR for the event under consideration, based on very simplified analyses and consideration of the observed heatup and pressurization rates.)

For event tree branches with the head on but with short-term makeup unavailable, the time to reach the high pressure isolation setpoint is estimated to be approximately six h. This estimate assumes all decay heat is absorbed in the coolant directly surrounding the core.

For event tree branches with the head detensioned, the time period to recover RHR is the time to reach boiling. This time period was 2.3 h for the loss of RHR at Susquehanna. For sequence 125, which involves a failure to recover RHR prior to boiling without an injection source and with the head detensioned, the time period would be even less.

For this event, the time to restore the faulted RPS bus (which caused the RHR isolation) was two hours. Assuming that

- the likelihood of not repairing the faulted bus as a function of time can be described as an exponential,

- no repair was possible during the first 20 min (to account for required response and diagnosis outside the control room),

- an additional 0.5-1.0 h is required to restart the RHR system once repaired (0.5 h if RHR venting is not required and 1.0 h if venting must be performed prior to restart), and

- the two-hour time-to-restore the RPS bus represents the median of repair times for this event,

the likelihood of failing to repair the bus can be represented by

$$P_{NREC\ BUS} = e^{-.415(t-.33)}, t \geq .33.$$

Skewing this an additional one-half hour to account for restoration of RHR results in an overall estimate of failing to recover RHR of

$$P_{NREC\ RHR} = e^{-.415(t-.83)}, t \geq .83.$$

For $t < .83$ h, $P_{NREC\ RHR} = 1.0$.

Applying this formula to the time periods discussed above, and subtracting the period of time that RHR was unavailable prior to the loss of the RPS bus (1.5 h), results in the following estimates for the probability of failing to recover RHR:

Sequence	Time to Recover RHR*	Probability
Head tensioned with short-term injection flow available (sequences 101-113)	6.0 h	0.12
Head tensioned with short-term injection flow unavailable (sequences 114-115)	4.5 h	0.22
Head detensioned but on, short-term injection flow available (sequences 116-123)	0.8 h	1.0
Head detensioned but on, short-term injection flow unavailable (sequences 124-125)	<0.8 h	1.0

---

*from discovery of loss of RPS bus

Main Condenser Available. The main condenser is modeled as a heat removal mechanism for sequences in which the condensate system is used as an injection source and the head is tensioned. The probability of the condenser being available for heat removal, given the condensate system is available, was assumed to be 0.5. The actual likelihood is dependant on the nature of the outage.

Required SRVs Opened. Sixteen SRVs are installed on Susquehanna. For sequences with the head tensioned (sequences 102-104, 106-108, and 111-113), opening of one or more SRVs provides success. For sequences with the head detensioned but still on the vessel (sequences 117-119) opening of three SRVs is required for success. In either case, failure of the valves to operate is dominated by dependant failure effects.

A probability of 1.6×10^{-4} was used for failure of multiple SRVs to open. This value was based on the observation of no such failures in the 1984-1990 time period, combined with a non-recovery likelihood of 0.12. This approach is consistent with the approach used to estimate this probability for other ASP evaluations, but includes a longer observation period and a lower probability of failing to recover to account for the 4-6 h typically available to open the valves [a non-recovery value of 0.71 is used for the probability of not recovering an ADS actuation failure in a one-half hour time period (see NUREG/CR-4674, Vol. 6) — this value was also used to estimate the likelihood of SRV failure for sequences with the head detensioned but on, since time periods for these sequences are short].

A value of 1.6×10^{-4} is consistent with failure probabilities which can be estimated from individual valve failure probabilities and beta factors, as described in NUREG/CR-4550, Vol 1, Rev. 1, "Analysis of Core Damage Frequency: Internal Events Methodology," and the conditional probability screening values used in the ASP program. The failure probabilities estimated using either approach are probably conservative, considering the number of valves potentially available for use. (NUREG/CR-4550, Vol 4, Rev. 1, Part 1, "Analysis of Core Damage Frequency: Peach Bottom, Unit 2, Internal Events," used a value of 1.0×10^{-6} for common cause SRV hardware faults, based on engineering judgement.)

Suppression Pool Cooling (Long-Term). On Susquehanna, like most BWRs, suppression pool cooling is a mode of RHR. One or more LPCI/RHR pumps take suction from the suppression pool, pump water through an RHR heat exchanger, and return it to the suppression pool. The suppression pool cooling mode of RHR consists of two redundant trains, each of which includes two parallel LPCI/RHR pumps, one heat exchanger, and two series return valves which must be opened to return flow to the suppression pool. For the train providing RHR prior to its loss, the suppression pool suction valves (normally open for LPCI but closed for RHR) must also be opened to provide suction to their respective pumps. During this event, RHR loop A was providing shutdown cooling, and hence opening of suction valves RHR 4A and 4C is assumed to be required.

Assuming availability of RHR service water and electric power, the equation for unavailability of suppression pool cooling is:

((RHR-4A+RHR-P1A)*(RHR-4C+RHR-P1C)+RHR-26A+RHR-24A*RHR-27A)
*(RHR-P1B*RHR-P1D+RHR-26B+RHR-24B*RHR-27B).

The minimal cutsets for this equation are

RHR-4A	RHR-4C	RHR-P1B	RHR-P1D
RHR-4A	RHR-4C	RHR-26B	
RHR-4A	RHR-4C	RHR-24B	RHR-27B
RHR-4A	RHR-P1C	RHR-P1B	RHR-P1D
RHR-4A	RHR-P1C	RHR-26B	
RHR-4A	RHR-P1C	RHR-24B	RHR-27B
RHR-P1A	RHR-4C	RHR-P1B	RHR-P1D
RHR-P1A	RHR-4C	RHR-26B	
RHR-P1A	RHR-4C	RHR-24B	RHR-27B
RHR-P1A	RHR-P1C	RHR-P1B	RHR-P1D
RHR-P1A	RHR-P1C	RHR-26B	
RHR-P1A	RHR-P1C	RHR-24B	RHR-27B
RHR-P1B	RHR-P1D	RHR-26A	
RHR-26A	RHR-26B		
RHR-26A	RHR-24B	RHR-27B	
RHR-P1B	RHR-P1D	RHR-24A	RHR-27A
RHR-26B	RHR-24A	RHR-27A	
RHR-24A	RHR-27A	RHR-24B	RHR-27B

Applying screening probabilities of 0.01 for failure of a motor-driven pump, 0.34 for failure to recover a faulted pump, 0.0001 for failure of a closed valve to open (because of the length of time available for recover, the NUREG-1150 value for a failure of a manual valve to open was employed), and 0.1, 0.3, and 0.5 for the conditional probabilities of the second, third, and fourth similar components to operate, results in an overall system failure probability estimate of 6.3×10^{-5}.

If only one train is available (because of maintenance on the other division), then the suppression pooling cooling failure probability is estimated to be 4.2×10^{-4}.

It should be noted that, because of the length of time available to recover suppression pool cooling (greater than 24 h), and the general lack of understanding of the reliability of such actions, this estimate has a high degree of uncertainty associated with it.

Analysis Results

Branch probabilities developed above were applied to the event tree model shown in Fig. 1 to estimated a conditional probability of subsequent severe core damage for the loss of RHR at Susquehanna. This conditional probability is 2.7×10^{-5}. Branch and selected sequence probabilities are shown in Fig. 2. Because of the way the event tree was constructed, the dominant sequences are associated with LPCI or low-pressure core spray (LPCS) success in providing RPV makeup. In the actual event, CRD flow was used for RPV makeup, and LPCI and LPCS were not actuated. The two dominant sequences both involve successful RPV makeup, failure to recover RHR (SDC) in the short-term, and failure to implement alternate core cooling because of failure to open at least one SRV (sequence 104) or failure to initiate suppression pool cooling (sequence 103). As discussed under ASP Modeling Approach and Assumptions: Branch Probabilities, above, the failure probabilities for these two branches are dependant on the probability of the branch failing when initially demanded and the probability of not restoring an initially failed branch over a period of perhaps 6-24 h. While the probability of initial failure on demand can be reasonably estimated, no information exists which would allow confident estimates of the probability of not recovering an initially failed component.

Additional calculations were performed to illustrate the sensitivity of the estimated conditional probability to analysis assumptions, as shown below:

Analysis Change	Conditional Probability
Probability of failing to open required SRVs = 1.0×10^{-6}	7.6×10^{-6}
Event could occur with head on, detensioned but on, or off [probabilities of each case specified under ASP Modeling Approach and Assumptions: Branch Probabilities (Head Status)]	5.8×10^{-5} (The dominant sequence for this case involves failure of RHR with the head on but detensioned, with failure to open at least three SRVs in the short-term.)
Random head status and one division out of service for maintenance and assumed non-recoverable	1.9×10^{-4} (The dominant sequence for this case also involves the head on but detensioned.)
Use of MSIV bypass valves/main condenser and HPCI for decay heat removal. (These decay heat removal methods are not addressed in ON-149-001.)	$\sim 4.8 \times 10^{-6}$

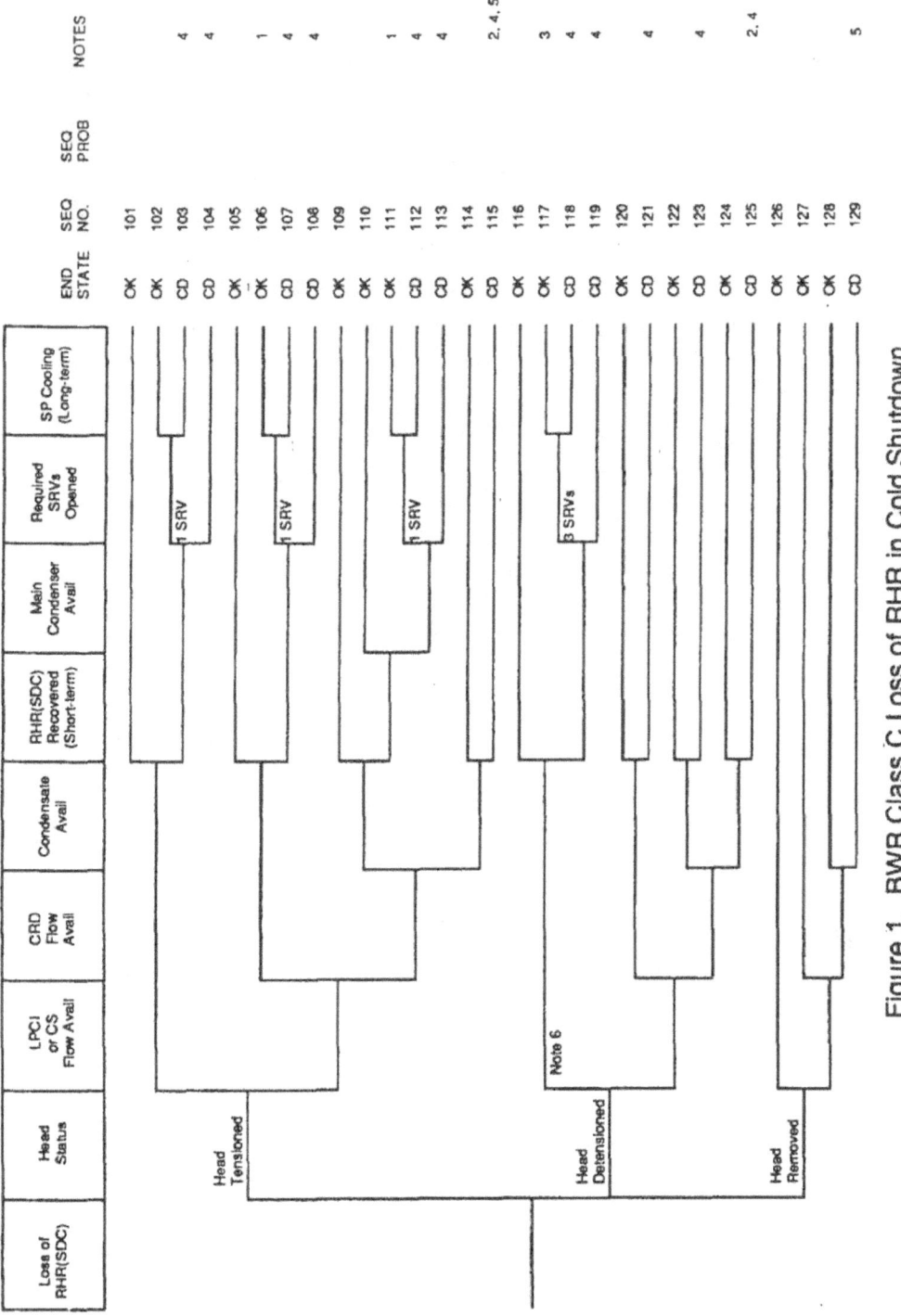

Figure 1. BWR Class C Loss of RHR in Cold Shutdown

Notes: 1. Suppression pool level will increase in this sequence.
2. Reduced time to recover RHR if recirculation pump unavailable since makeup required to achieve natural circulation is also unavailable.
3. Water in main steam lines may overstress these lines.
4. Use of RWCU/Condensate Transfer to transfer hot water to the condenser or condensate storage tank will increase the time available to recover RHR or initiate suppression pool cooling.
5. Alternate injection sources such as service water may also provide injection.
6. If primary and secondary containment cannot be established, this sequence is prescribed.

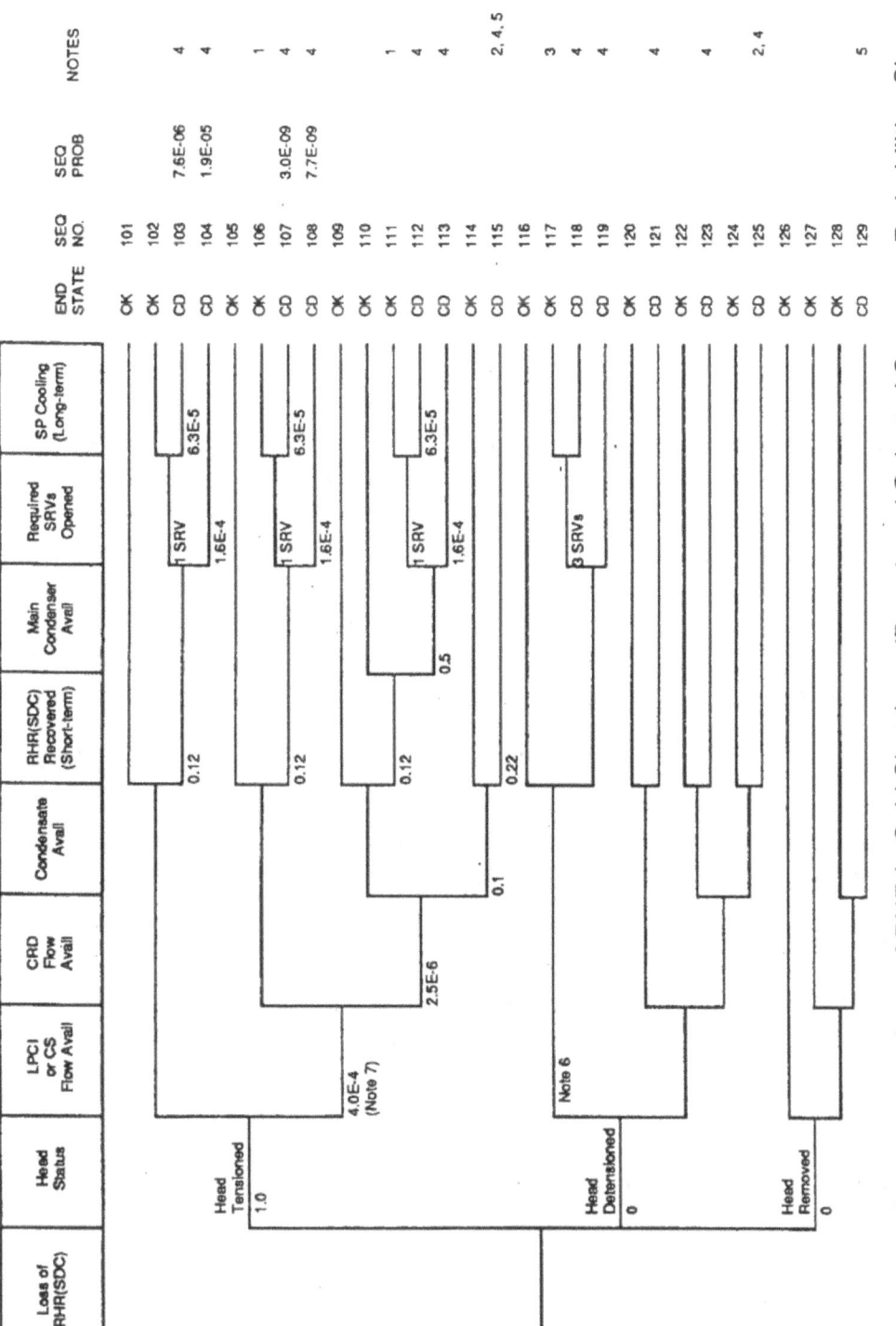

Figure 2. Susquehanna Loss of RHR in Cold Shutdown (Branch and Selected Sequence Probabilities Shown)

Notes: 1. Suppression pool level will increase in this sequence.
2. Reduced time to recover RHR if recirculation pump unavailable since makeup required to achieve natural circulation is also unavailable.
3. Water in main steam lines may overstress these lines.
4. Use of RWCU/Condensate Transfer to transfer hot water to the condenser or condensate storage tank will increase the time available to recover RHR or initiate suppression pool cooling.
5. Alternate injection sources such as service water may also provide injection.
6. If primary and secondary containment cannot be established, this sequence is prescribed.
7. LPCS failure probability.

ACCIDENT SEQUENCE PRECURSOR PROGRAM COLD SHUTDOWN EVENT ANALYSIS

LER No.:   397/88-011
Event Description: Reactor cavity draindown
Date of Event: May 1, 1988
Plant: Washington Nuclear Plant 2

Summary

Washington Nuclear Plant 2 (WNP 2) was at cold shutdown on May 1, 1988. While changing from loop "B" to loop "A" of residual heat removal/shutdown cooling (RHR/SDC), the operator inadvertently opened the suppression pool suction valve on loop "B" before the reactor RHR/SDC suction valve on loop "B" was fully closed. The two valves were simultaneously open for approximately 40 sec which provided a drain path for the reactor pressure vessel (RPV) to the suppression pool. The RPV water level dropped fast enough to cause a low level scram and isolation of RHR/SDC. The RHR/SDC isolation stopped the RPV level drop, but RHR/SDC was lost for about seven min until level was restored and the isolation was reset. The conditional probability of subsequent severe core damage estimated for the event is 4.6×10^{-5}. Dominant sequences are associated with failure to implement alternate core cooling strategies in the event that RHR could not be recovered in the short term. The calculated probability is strongly influenced by estimates of the likelihood of failing to recover initially faulted systems over time periods of 6-24 h. These estimates involve substantial uncertainty, and hence the overall core damage probability estimated for the event also involves substantial uncertainty.

Event Description

On May 1, 1988 WNP 2 was at cold shutdown with the reactor coolant temperature between 140F and 160F. RHR "B" was on line in the SDC mode, RHR "A" was in standby, lined up for emergency core cooling system (ECCS) actuation, and reactor recirculation pump 1A was operating at 15 cycles per second. The plant had begun a refueling outage on April 29, 1988 and operators were preparing to changeover to loop "A" of RHR for SDC and to place loop "B" of RHR in standby for ECCS actuation. The procedure governing this evolution required the operator to close the reactor suction valve for SDC (RHR-6B) before he opened the suppression pool suction valve (RHR-4B) when he placed loop "B" in standby. However, the operator did not wait for RHR-6B to fully close before opening RHR-4B. This action violated the approved operating procedure as well as a "permanent operator aid" caution label on the control panel. Both these valves have stroke times of about 120 sec, and, as a result, both valves were simultaneously open for approximately 40 sec. This was long enough for the reactor cavity to gravity drain about 10,000 gal of water to the suppression pool. The draindown was stopped when the reactor water level reached the RPV low level scram and SDC isolated. The isolation signal closed the SDC suction isolation valves inside primary containment (RHR-8 and -9), but closing RHR-8 and -9 also failed RHR SDC. The operator backed-up the automatic isolation by manually closing RHR-8 and -9. RPV water level was restored in about seven min using the control rod drive (CRD) and condensate systems and SDC was reestablished at that time.

Fig. 1 is a diagram of loop B of the RHR system for this plant.

Additional Event-Related Information

Reactor scram and the automatic isolation of RHR/SDC from the reactor recirculation system occur at 174 in above the top of active fuel (TAF). The high-pressure core spray (HPCS) system automatically lines up for and initiates vessel makeup and the reactor recirculation pumps trip off at 111" above TAF. LPCI and LPCS initiation occurs at 32" above TAF. At this point, RHR automatically lines up for and initiates low-pressure coolant injection (LPCI) mode. That is, appropriate valves line up for pump suction on the suppression chamber, SDC isolation, and test return isolation. Also, the low-pressure core spray (LPCS) system automatically lines up for and initiates vessel makeup.

A previous event (LER 397/85-030) that was referred to in the LER occurred in 1985. That event was remarkably similar to this event except in the 1985 incident the operator waited 30 sec before he began opening the suppression pool suction valve. Consequently, the level did not drop as far as in this event. SDC was lost for about one h; however, the plant had been shutdown for approximately four d for an extended maintenance outage following a run for over three weeks at a reduced power of 45%.

ASP Modeling Approach and Assumptions

Event Tree for Loss of RPV Inventory

An event tree model of sequences to core damage given the loss of RPV inventory is shown in Fig. 2. If RHR isolation successfully terminates the inventory loss, the event tree describes sequences associated with loss of SDC. This portion of the event tree was developed based on procedures (e.g. Procedure PPM 2.4.2, "RHR System", September 7, 1990) in effect at WNP 2 at the time of the event, the Plant Technical Specifications, and the Final Safety Analysis Report (FSAR). If RHR isolation fails, the event tree describes the use of LPCI, core spray, or HPCS (break-size dependant), plus long-term suppression pool cooling to mitigate core damage.

The following comments are applicable to this event tree:

a. Core damage end state. Core damage is defined for the purpose of this model as reduction in RPV level above TAF or failure to remove heat from the suppression pool in the long term. With respect to RPV inventory, this definition may be conservative, since steam cooling may limit clad temperature increase in some situations. However, choice of TAF as the damage criterion allows the use of simplified calculations to estimate the time to an unacceptable end state.

b. Short-term recovery of RHR. All historic losses of RHR have been recovered before RPV level would have dropped to below TAF. Including RHR recovery allows operational events to be more realistically mapped onto the event tree model. Short-term RHR recovery can be delayed if a recirculation pump can be started or if RPV level can be raised to permit natural circulation. Availability of RPV injection to raise water level for natural circulation is included in the model.

c. Successful termination of the loss of RHR is defined as recovery of RHR or provision of alternate decay heat removal via the suppression pool or main condenser, or, if the head is removed, via refueling cavity boiling. Short-term decay heat removal methods (such as feed with bleed to a tank) with subsequent long-term recovery of RHR, is not addressed in the event tree, although such an approach can provide additional time to implement a long-term core cooling approach.

d. Three pressure vessel head states are addressed in the event tree: head on and tensioned, head on and detensioned, and head off. If the head is on and tensioned, then decay heat removal methods which require pressurization are assumed to be viable. If the head is on, but detensioned, then failure to maintain the RPV depressurized is also assumed to proceed to core damage (this assumption is conservative). If the head is off, then makeup at a rate equal to boil-off is assumed to provide core cooling.

e. Five makeup sources are shown on the event tree: LPCI, LPCS, HPCS, CRD flow and the condensate system. Branches for these sources are shown before short-term RHR recovery. This is because injection from any source to raise RPV level and allow natural circulation substantially increases the amount of time available for recovery of RHR. The five makeup sources have been placed before RHR recovery to address this issue, even though the need for significant flow from these systems is only required if RHR is not recovered.

If RHR isolation fails, RPV makeup must compensate for the flow from the RHR system to the suppression pool. Sources of this makeup must take suction from the suppression pool to prevent the suppression pool from being completely filled. The use of LPCI, LPCS, or HPCS is included on the event tree.

f. In the event that RHR cannot be recovered, then alternate core cooling sequences are included in the event tree. Based on studies done at Susquehanna, if the head is tensioned, these involve allowing the RPV to repressurize, opening of at least one safety relief valve (SRV), and dumping decay heat to the suppression pool. If the condenser and condensate system are available, then decay heat can also be dumped to the condenser. If the head is detensioned, then decay heat must be removed without the RPV being pressurized. Again, based on studies done at Susquehanna, this requires opening of at least three SRVs and recirculating water to the suppression pool using the LPCS or LPCI pumps. For all cooling modes involving the suppression pool, suppression pool cooling must be initiated in sufficient time to prevent the suppression pool from exceeding its temperature limit. If the head is removed, then any makeup source greater than ~200 gpm, combined with boiling in the RPV, will provide adequate core cooling.

Fig. 2 includes the following core damage sequences:

Sequence	Description

Sequences with the Head Tensioned and Loss of Inventory Terminated

104   Unavailability of long-term heat removal from the suppression pool with failure to recover RHR and unavailability of the main condenser but following successful alternate short-term decay heat removal using the condensate system with relief to the suppression pool via one or more SRVs.

105   Failure to recover RHR and unavailability of the main condenser and failure to initiate alternate short-term decay heat removal due to unavailability of the SRVs for relief to the suppression pool.

108   Unavailability of long-term heat removal from the suppression pool with failure to recover RHR but following successful alternate short-term decay heat removal using LPCI or LPCS injection and relief to the suppression pool via one or more SRVs.

109   Failure to recover RHR and failure to initiate alternate short-term decay heat removal due to unavailability of the SRVs for relief to the suppression pool.

112   Similar to sequence 108 except the condensate system, LPCI, and LPCS are unavailable. RPV injection provided using HPCS flow.

113   Similar to sequence 109 except the condensate system, LPCI, and LPCS are unavailable. RPV injection provided using HPCS flow.

116   Similar to sequence 108 except LPCI, LPCS, and HPCS are unavailable. RPV injection provided using CRD flow.

117   Similar to sequence 109 except LPCI, LPCS, and HPCS are unavailable. RPV injection provided using CRD flow.

119   Failure to recover RHR and unavailability of LPCI, LPCS, HPCS, CRD flow and the condensate system to raise RPV level to provide for natural circulation. The time available to recover RHR in this sequence is less than for sequences with RPV injection unless a recirculation pump can be started, since RPV level cannot be raised to provide for natural circulation cooling.

Sequences with the Head Detensioned and Loss of Inventory Terminated

122   Unavailability of long-term heat removal from the suppression pool with failure to recover RHR but with successful alternate decay heat removal using LPCI or LPCS injection with discharge to the suppression pool using three or more SRVs.

123   Failure to recover RHR and failure to initiate alternate short-term decay heat removal due to unavailability of three or more SRVs for relief to the suppression pool.

125   Failure to recover RHR with unavailability of LPCI and LPCS for alternate decay heat removal. HPCS flow provides sufficient water to raise RPV level and allow natural circulation, extending the time available to recover RHR.

127   Failure to recover RHR with unavailability of LPCI and LPCS for alternate decay heat removal. HPCS flow is unavailable but CRD flow provides sufficient water to raise RPV level and allow natural circulation, extending the time available to recover RHR.

129   Similar to sequence 127 except CRD flow is also unavailable. Condensate is used to increase RPV level and allow natural circulation.

Sequence with the Head Removed and Loss of Inventory Terminated

134   Unavailability of LPCI, LPCS, HPCS, CRD flow and condensate for RPV makeup. Core damage in the long term if a supplemental makeup source cannot be provided.

Sequences without Termination of Inventory Loss

136   Unavailability of long term decay heat removal from the suppression pool with successful LPCI or LPCS injection to make up for the loss of RPV inventory.

138   Similar to sequence 138 except LPCI and LPCS are unavailable. HPCS (with suction from the suppression pool) provides injection. HPCS injection success is break-size dependant.

139   Unavailability of LPCI, LPCS, and HPCS to provide makeup for the loss of RPV inventory.

Branch Probabilities

<u>Loss of Inventory Terminated by RHR ISO</u>. Closure of either RHR-8 or RHR-9 at the SDC isolation setpoint will isolate RPV flow to the suppression pool. Assuming a screening probability of 0.01 for the failure of a motor-operated valve to close and 0.1 for the conditional probability of

the second valve results in a branch failure probability estimate of 1.0×10^{-3}. Note that closure of RHR-6B would also terminated the RPV inventory loss. This valve was not considered in estimating the failure probability for this branch.

Head Status. A review of WNP 2 refueling outages over the last five and one half years indicates an average outage duration of 75.6 d. Assuming that two days of the outage are not at cold shutdown, and that the total time during an outage that the head is on but detensioned is approximately two days, results in the following time periods for the three head states over a period: head on, 4 d; head detensioned but on, 2 d; and head off, 67.6 d.

In addition to refueling outages, there has been 47 outages of an average length of 4.6 d. If we again assume two days per outage not at cold shutdown, and assume that during the remainder of the time the plant is at cold shutdown with the head on, the following overall fractions of time for the three head states are estimated:

head on	0.27
head on but detensioned	0.02
head off	0.71

Condensate Available. While the condensate pumps can provide more than adequate makeup, they are often unavailable during a refueling outage because of work on the secondary system. However, the condensate system was available during this event and was used to restore the RPV level following the reactor cavity draindown. A failure probability of 0.01 was assumed.

LPCI or CS Flow Available. For sequences involving successful RHR isolation, flow from any LPCI or LPCS pump will provide adequate makeup. To simplify the estimation of the probability of failure of suppression pool cooling (which is dependant on the status of the LPCI trains which also provide SDC), only the failures associated with LPCS and the non-RHR train of LPCI were used to estimate this branch probability. For WNP 2, LPCS consists of one train. The train includes one pump with a single, normally open motor-operated suction valve and a single normally-closed discharge (RPV injection) valve. The pump suction source is normally the suppression pool. LPCI train C consists of a motor-driven pump, a normally-open motor-operated suction valve and a normally-closed motor-operated discharge (RPV injection) valve. The pump suction source is also the suppression pool. Assuming that normally-open valves and check valves do not contribute substantially to system unavailability, the equation for failure of LPCS is therefore

$$(LPCS-P1 + LPCS-5) * (RHP-P2C + RHR-42C)$$

Applying screening probabilities of 0.01 for failure of a motor-driven pump to start and run and failure of a motor-operated valve to open, 0.1 for the conditional probability of the second similar component to operate, and a likelihood of 0.34 of not recovering a failed LPCI train or core spray system in the short-term results in an overall system failure probability estimate for this branch of 7.5×10^{-4}.

For sequences involving failure to isolate RHR, two of the four LPCI and LPCS trains must operate to provide makeup for the flow path to the suppression pool. The operating RHR train's suction supply must be aligned to the suppression pool. In the two non-operating LPCI trains and the LPCS train, the pumps must start and the discharge isolation valves must open. Since success requires two of four trains, three of four trains must fail for injection failure:

$$(LPCS\text{-}P1 + LPCS\text{-}5) * (RHR\text{-}P2C + RHR\text{-}42C) * (RHR\text{-}PA2 + RHR\text{-}42A * RHR\text{-}53A) +$$
$$(LPCS\text{-}P1 + LPCS\text{-}5) * (RHR\text{-}P2C + RHR\text{-}42C) * (RHR\text{-}6B + RHR\text{-}4B) +$$
$$(LPCS\text{-}P1 + LPCS\text{-}5) * (RHR\text{-}PA2 + RHR\text{-}42A * RHR\text{-}53A) * (RHR\text{-}6B + RHR\text{-}4B) +$$
$$(RHR\text{-}P2C + RHR\text{-}42C) * (RHR\text{-}PA2 + RHR\text{-}42A * RHR\text{-}53A) * (RHR\text{-}6B + RHR\text{-}4B)$$

The minimal cutsets for this equation are

RHR-4B	RHR-P2A	RHR-P2C
RHR-42C	RHR-6B	RHR-P2A
RHR-6B	RHR-P2A	RHR-P2C
LPCS-5	RHR-42C	RHR-P2A
LPCS-P1	RHR-42C	RHR-P2A
LPCS-5	RHR-P2A	RHR-P2C
LPCS-P1	RHR-P2A	RHR-P2C
LPCS-5	RHR-42C	RHR-4B
LPCS-P1	RHR-42C	RHR-4B
LPCS-5	RHR-42C	RHR-6B
LPCS-P1	RHR-42C	RHR-6B
LPCS-5	RHR-4B	RHR-P2C
LPCS-P1	RHR-4B	RHR-P2C
LPCS-5	RHR-6B	RHR-P2C
LPCS-P1	RHR-6B	RHR-P2C
LPCS-5	RHR-4B	RHR-P2A
LPCS-P1	RHR-4B	RHR-P2A
LPCS-5	RHR-6B	RHR-P2A
LPCS-P1	RHR-6B	RHR-P2A
RHR-42C	RHR-4B	RHR-P2A

Applying the screening probabilities described above results in a branch probability estimate of 5.6×10^{-5}.

HPCS Flow Available. HPCS at WNP 2 consists of one train. This train includes one pump with a single, normally open motor-operated suction valve and a single normally-closed discharge (RPV injection) valve. The pump suction source for HPCS is normally the condensate storage tank (CST). Again assuming that normally-open valves and check valves do not contribute substantially to system unavailability, the equation for failure of HPCS is therefore

$$HPCS-P1 + HPCS-4$$

Applying the screening probabilities described above results in an overall system failure probability estimate for HPCS of 6.8×10^{-3}.

For sequences involving failure to isolate RHR, HPCS cannot provide makeup for flow from the open suction valve. The unavailability of HPCS for those sequences is 1.0.

CRD Flow Available. At cold shutdown pressures, one of two CRD pumps can provide makeup. Since one pump is typically running, the system will fail if that pump fails to run and if the other (standby) pump fails to start and run. Assuming a probability of 0.01 for failure of the standby CRD pump to start, and 3.0×10^{-5}/hr for failure of a pump to run, results in an estimated failure probability for CRD flow of 2.5×10^{-6}. In this estimate, a short-term non-recovery likelihood of 0.34 was applied to the non-running pump failure-to-start probability, consistent with the approach used to estimate the failure probability for the core spray system. A mission time of 24 h was also assumed.

If only one train is available (because of maintenance on the opposite division), then the CRD failure probability is estimated to be 7.2×10^{-4}.

RHR (SDC) Recovered (Short-Term). For WNP 2, RHR can be restored to service provided RPV level is greater than the low-level isolation level and RPV pressure is less than the high pressure isolation pressure, and, of course, the cause of the initial loss of RHR is repaired.

For event tree branches with the head on and for which reactor vessel (RV) inventory was increased to provide for natural circulation, RHR must be recovered prior to RV pressure reaching the high pressure isolation setpoint (135 psig at WNP 2), which would prevent opening the suction line isolation valves and restoring RHR. Once the high-pressure isolation setpoint is reached, operation of at least one SRV is assumed to be required, based on the studies done at Susquehanna, and the sequence proceeds with RPV depressurization and the use of RHR in the suppression pool cooling mode to remove decay heat. In estimating the probability of not recovering RHR (SDC), the time period of concern for these sequences is from initial loss of RHR until the high-pressure isolation setpoint is reached. (Approximately 7.5 h from the loss of RHR for the event under consideration, based on very simplified analyses and consideration of the observed heatup and pressurization rates.)

For event tree branches with the head on but with short-term makeup unavailable, the time to reach the high pressure isolation setpoint is estimated to be approximately six hours. This estimate assumes all decay heat is absorbed in the coolant directly surrounding the core.

For event tree branches with the head detensioned, the time period to recover RHR is the time to reach boiling. The time to reach boiling following the loss of RHR at WNP 2 was approximately 1 h. For sequence 131, which involves a failure to recover RHR prior to boiling without an injection source and with the head detensioned, the time period would be even less.

For this event, the time to restore RHR(SDC) was about seven minutes when the vessel level was recovered and the isolation was reset.

This event involved no actual component failures or any loss of supplied power. The plant was also at operational condition 4, which means ECCS was available and operable. Therefore, the probability of failing to recover RHR was assumed to be dictated by the failure probabilities of components in the LPCI system. No additional impact resulting from human error was assumed.

Failure to recover RHR is dominated by failure of either RHR-8 or RHR-9 to open, both RHR pumps to start, or both injection valves to open. Applying the screening probabilities described above results in a branch probability estimate of 7.5×10^{-3}.

Main Condenser Available. The main condenser is modeled as a heat removal mechanism for sequences in which the condensate system is used as an injection source and the head is tensioned. The probability of the condenser being available for heat removal, given the condensate system is available, was assumed to be 0.5. The actual likelihood is dependant on the nature of the outage.

Required SRVs Opened. Eighteen SRVs are installed at WNP 2. The following analysis is based on the studies done at Susquehanna. For sequences with the head tensioned (sequences 102-104, 106-108, 110-112, and 115-117), opening of one or more SRVs provides success. For sequences with the head detensioned but still on the vessel (sequences 121-123) opening of three SRVs is required for success. In either case, failure of the valves to operate is dominated by dependant failure effects.

A probability of 1.6×10^{-4} was used for failure of multiple SRVs to open. This value was based on the observation of no such failures in the 1984-1990 time period, combined with a non-recovery likelihood of 0.12. This approach is consistent with the approach used to estimate this probability for other ASP evaluations, but includes a longer observation period and a lower probability of failing to recover to account for the 4-6 h typically available to open the valves [a non-recovery value of 0.71 is used for the probability of not recovering an ADS actuation failure in a one-half hour time period (see NUREG/CR-4674, Vol. 6). This value was also used to estimate the likelihood of SRV failure for sequences with the head detensioned but on, since time periods for these sequences are short].

A value of 1.6×10^{-4} is consistent with failure probabilities which can be estimated from individual valve failure probabilities and beta factors, as described in NUREG/CR-4550, Vol 1, Rev. 1, "Analysis of Core Damage Frequency: Internal Events Methodology," and the conditional probability screening values used in the ASP program. The failure probabilities estimated using either approach are probably conservative, considering the number of valves potentially available for use. (NUREG/CR-4550, Vol 4, Rev. 1, Part 1, "Analysis of Core Damage Frequency: Peach Bottom, Unit 2, Internal Events," used a value of 1.0×10^{-6} for common cause SRV hardware faults, based on engineering judgement.)

Suppression Pool Cooling (Long-Term). At WNP 2, like most BWRs, suppression pool cooling is a mode of LPCI. The LPCI system consists of three independent loops at WNP 2, and each loop contains its own motor-driven pump, has a suction from the suppression pool, and is capable of discharging water to the reactor vessel via a separate nozzle or back to the suppression pool via a full-flow test line. Two of these loops have a heat exchanger which is cooled by normal or standby service water. The suppression pool cooling mode of RHR consists of two redundant trains, each of which includes an RHR/LPCI pump, a heat exchanger, and a single return valve which must be opened to return flow to the suppression pool. For the train providing RHR (SDC), the suppression pool suction valve (normally open for LPCI but closed for RHR-SDC) must also be opened to provide suction to its respective pump. During this event, RHR loop A had been providing shutdown cooling and RHR loop B was just going into standby. It was conservatively assumed opening of suction valve RHR-V-4A was required for this mode of operation.

Assuming availability of RHR service water and electric power, the equation for unavailability of suppression pool cooling is:

$$(RHR\text{-}4A + RHR\text{-}P2A + RHR\text{-}24A) * (RHR\text{-}4B + RHR\text{-}P2B + RHR\text{-}24B)$$

The minimal cutsets for this equation are

RHR-4A	RHR-4B
RHR-4A	RHR-P2B
RHR-4A	RHR-24B
RHR-P2A	RHR-4B
RHR-P2A	RHR-P2B
RHR-P2A	RHR-24B
RHR-24A	RHR-4B
RHR-24A	RHR-P2B
RHR-24A	RHR-24B

Applying screening probabilities of 0.01 for failure of a motor-driven pump, 0.34 for failure to recover a faulted pump, 0.0001 for failure of a closed valve to open (because of the length of time available for recover, the NUREG-1150 value for a failure of a manual valve to open was employed), and 0.1 for the conditional probability of the second similar component to operate, results in an overall system failure probability estimate of 3.5×10^{-4}.

The conditional failure probability for suppression pool cooling given failure to recover RHR (SDC) in the short term is 4.5×10^{-2}. This value is influenced by the fact that failure of both RHR/LPCI pumps faults both branches. If only one train is available (because of maintenance on the other division), then the suppression pool cooling failure probability is estimated to be 3.6×10^{-3}.

For sequences involving a failure to terminate the loss of inventory with LPCI or LPCS success, a branch probability of 3.0×10^{-4} is estimated.

It should be noted that, because of the length of time available to recover suppression pool cooling (greater than 24 h), and the general lack of understanding of the reliability of such actions, this estimate has a high degree of uncertainty associated with it.

Analysis Results

Branch probabilities developed above were applied to the event tree model shown in Fig. 1 to estimate a conditional probability of subsequent severe core damage for the reactor cavity draindown at WNP 2. This conditional probability is 4.6×10^{-5}. The dominant sequences involve successful termination of the loss of inventory, successful RPV makeup, failure to recover RHR (SDC) in the short-term, unavailability of the main condenser for decay heat removal, and failure to implement alternate core cooling because of failure to open at least one SRV (sequence 105) or failure to initiate suppression pool cooling (sequence 104). As discussed under ASP Modeling Approach and Assumptions: Branch Probabilities, above, the failure probabilities for these two branches are dependant on the probability of the branch failing when initially demanded and the probability of not restoring an initially failed branch over a period of perhaps 6-24 h. While the probability of initial failure on demand can be reasonably estimated, no information exists which would allow confident estimates of the probability of not recovering an initially failed component over these time periods.

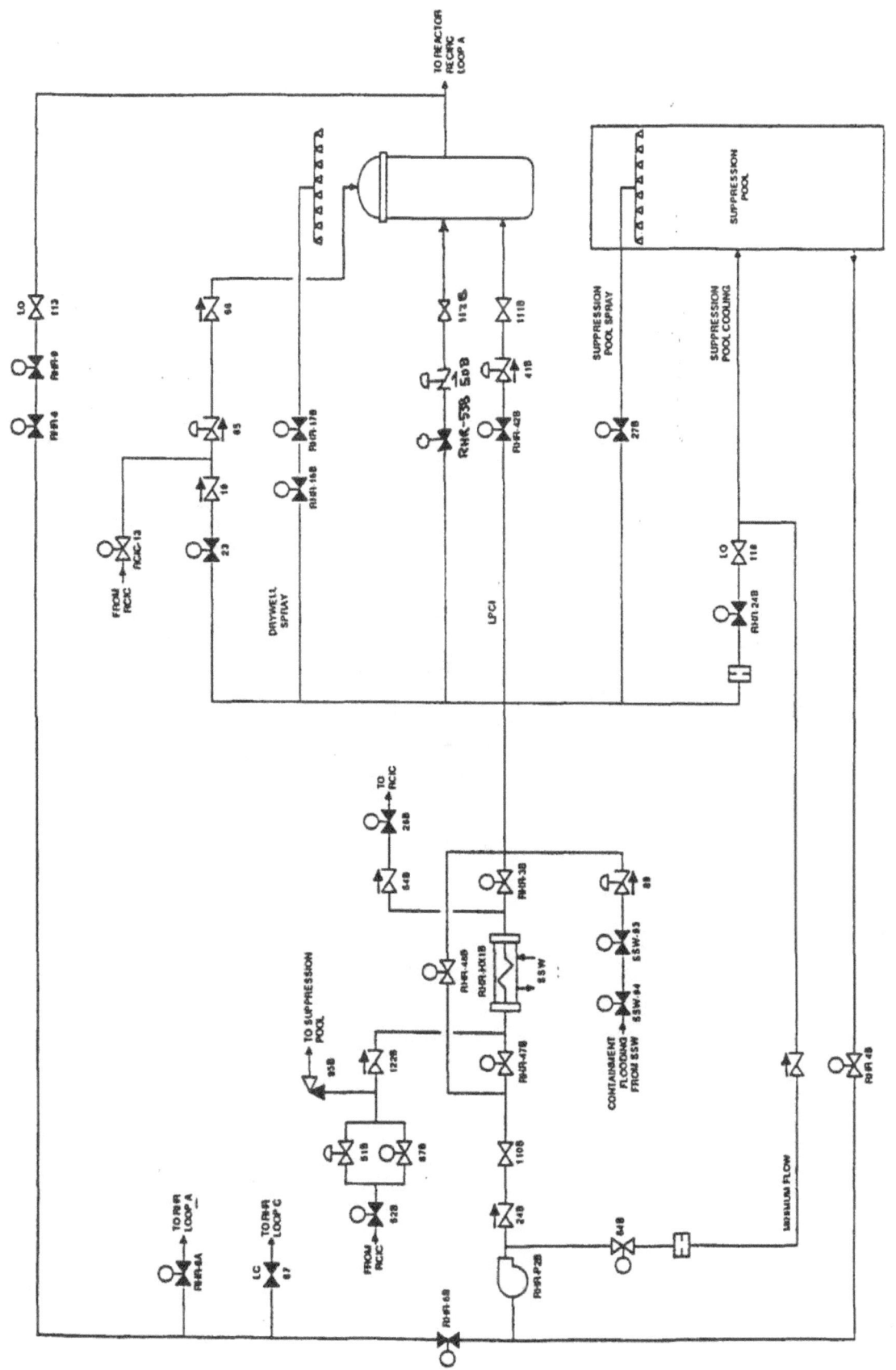

Fig. 1. WNP2 RHR system, loop B

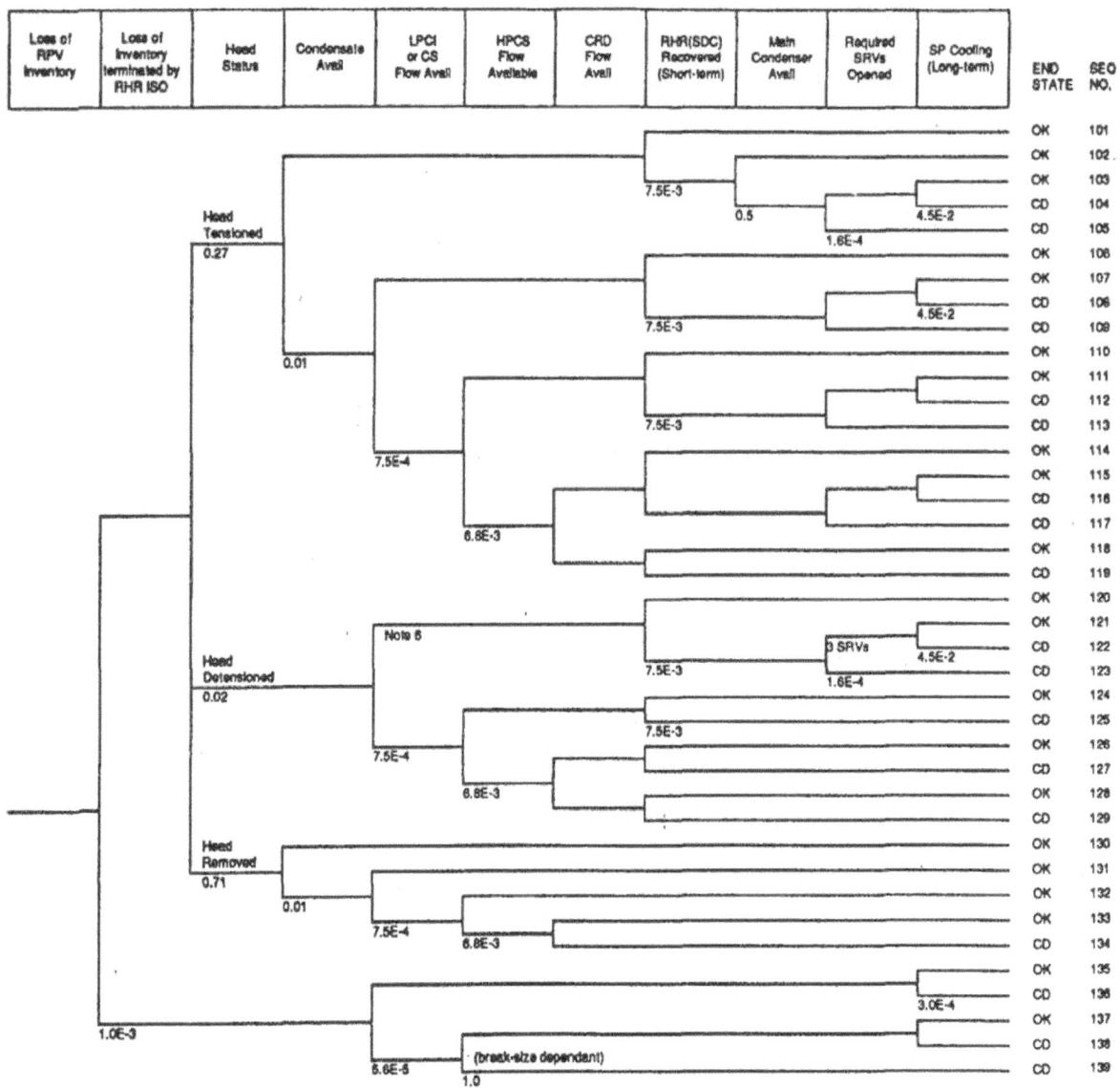

Fig. 2. Event Tree Model for LER 397/88-011

ACCIDENT SEQUENCE PRECURSOR PROGRAM COLD SHUTDOWN EVENT ANALYSIS

LER No: 456/89-016
Event Description: RHR suction relief valve drains 64,000 gal from RCS
Date of Event: December 1, 1989
Plant: Braidwood 1

Summary

A residual heat removal (RHR) pump suction relief valve opened below its design setpoint and would not reseat. Approximately 64,000 gal flowed through the relief valve to the boron recycle holdup tank before the leakage path was isolated. About 54,000 gal were made up from the refueling water storage tank (RWST). Identification of the faulted valve was delayed because the valve was in the non-operating RHR train, and initial operator response addressed the operating train. The event occurred after a full core reload, when no decay heat load existed, and hence the conditional probability of subsequent core damage is very small. Had the event occurred when decay heat removal was required, its conditional probability would still be below 1.0×10^{-6}.

Event Description

Prior to the event, Braidwood 1 was in cold shutdown with "A" RHR train in service. "B" train was aligned, but not operating. Reactor coolant pressure was 350 psig, and temperature was 170F. The pressurizer was solid, and preparations were under way to draw a steam bubble.

By 0142, reactor coolant system (RCS) pressure had risen to 404 psig when the 1B RHR pump suction relief valve opened. The pressure setpoint for this valve was supposed to be 450 psig. Inspection and testing after the event indicated an as-installed set pressure of approximately 410 psig (apparently because of incorrect maintenance 20 months earlier — April, 1988). In addition, the nozzle ring setting was out of adjustment by 233 notches, which prevented the valve from reclosing during the event.

Pressurizer level began declining from off-scale high and decreased rapidly. The operator began reducing letdown flow and increasing charging flow. Boron recycle holdup tank level began increasing rapidly. By 0151, pressurizer level was off-scale low. Operations concluded that a RHR pump suction relief valve had lifted and failed to reseat.

Initially, plant operators assumed that the RCS leakage was from the operating RHR train (valve RH 8708A). At 0155, "A" RHR train was removed from service and "B" train placed in service. The operating charging pump was aligned to the RWST. RCS pressure stabilized at 272 psig. The utility believes that the RCS level at this point was somewhere in the lower portion of the pressurizer surge line, and that, by this time, charging flow equaled leakage from the relief valve.

This elevation corresponds generally to the lower portion of the steam generator tubes and to the upper portion of the reactor vessel. Reactor vessel level instrumentation indicated 100% at all times, and subsequent RCS venting using the head vents indicated no gases in the reactor vessel.

Charging pump 1B breaker was racked-in and the pump was started at 0235. By 0245, pressurizer level indicated above 0%, and 1B charging pump was secured. Reactor pressure was 310 psig. By 0254, pressurizer level had again declined off-scale, and RCS pressure was declining. This implies that the leakage rate was greater than the capacity of the operating charging pump and that the lowest RCS level achieved may have been at 0235, just before charging pump 1B was first started. Charging pump 1B was restarted at 0254, and pressurizer level rose above 0% at 0302, whereupon charging flow from the two pumps was throttled. Holdup tank levels continued to increase.

At 0319, it was finally determined that the 1B RHR pump relief valve (RH 8708B) was leaking. By 0350, RHR train "A" was again in service and RHR train "B" was isolated, ending the event. Approximately 64,000 gal were lost through the RHR pump suction relief valve. About 54,000 gal were made up from the RWST.

A simplified drawing of the Braidwood RHR system is provided in Fig. 1. A detailed sequence of events is provided in Attachment A.

Additional Event-Related Information

Braidwood was in the 101st day of a refueling outage. A complete fuel reload was performed and the potential for temperature increase from decay heat did not exist. The RCS inventory was always sufficient to keep the core covered and no loss of shutdown cooling occurred.

As specified in attachment A, one centrifugal charging pump was operating prior to the event. The other charging pump was tagged out-of-service with its breaker racked out (as required by the plant Technical Specifications for this operating mode), as were both safety injection (SI) pumps. The tagged out charging pump was restored to service during the event, and the two SI could apparently also have been restored to service if required. All four steam generators (SGs) were available with water levels between 63 and 69 percent.

The Braidwood procedure for loss of RHR cooling applicable at the time of the event also addresses loss of RCS inventory while the RHR system is in operation. This procedure specifies a variety of methods to provide decay heat removal: bleed and feed using excess letdown and normal charging, steaming of intact SGs, bleed and feed using the pressurizer power-operated relief valves (PORVs) and normal charging, refuel cavity to fuel pool cooling, SI pump hot leg injection, accumulator injection, and gravity feed from the RWST. In addition, the procedure includes instructions for venting the RHR trains, including requirements to close the RHR drop line valves during venting. Had the open relief valve not been discovered, and the charging and SI systems and the accumulators failed to provide RCS makeup such that the RHR pumps had to be vented, then closure of the drop line valves would have isolated the open relief valve and terminated the event. At this point, the SGs could have been steamed to provide decay heat removal.

Analysis Approach

The analysis approach for this event depends upon when the relief valve could have lifted. For the actual event, the valve lifted after a complete fuel reload when there was no decay heat. In this case, the conditional probability of subsequent core damage is extremely small.

If the relief valve had lifted shortly after entering shutdown, then RCS makeup from the charging system, SI system or accumulators would have provided for extended decay heat removal until the open relief valve was found. Once the open valve was isolated and RCS inventory loss terminated, the SGs or intact RHR train could have been used for decay heat removal. For this situation, the following failures would have been required before core damage would have occurred: (1) failure to align the charging pumps to the RWST or failure to start the non-operating pump, (2) failure of both SI pumps to provide RCS injection, (3) failure of the operators to use the accumulators for RCS makeup, and (4) failure to close the RCS drop line valves or failure to use the SGs or intact RHR train for decay heat removal.

Applying typical ASP failure probabilities to components in the above systems results in a core damage probability estimate considerably below 1×10^{-6}. If one division had been out of service for maintenance, then only the operating RHR train drop valves would have been open. In this case, the operators would have rapidly identified the appropriate relief valve and terminated the loss of RCS inventory. Following this, the operating charging pump would have provided adequate decay heat removal until the other RHR train could be restored to service.

Analysis Results

Because a complete fuel reload was completed prior to this event and no decay heat load existed, the event is estimated to have a very small probability of subsequent core damage. Had the event occurred when decay heat removal was required, its conditional probability would still have been below 1.0×10^{-6}.

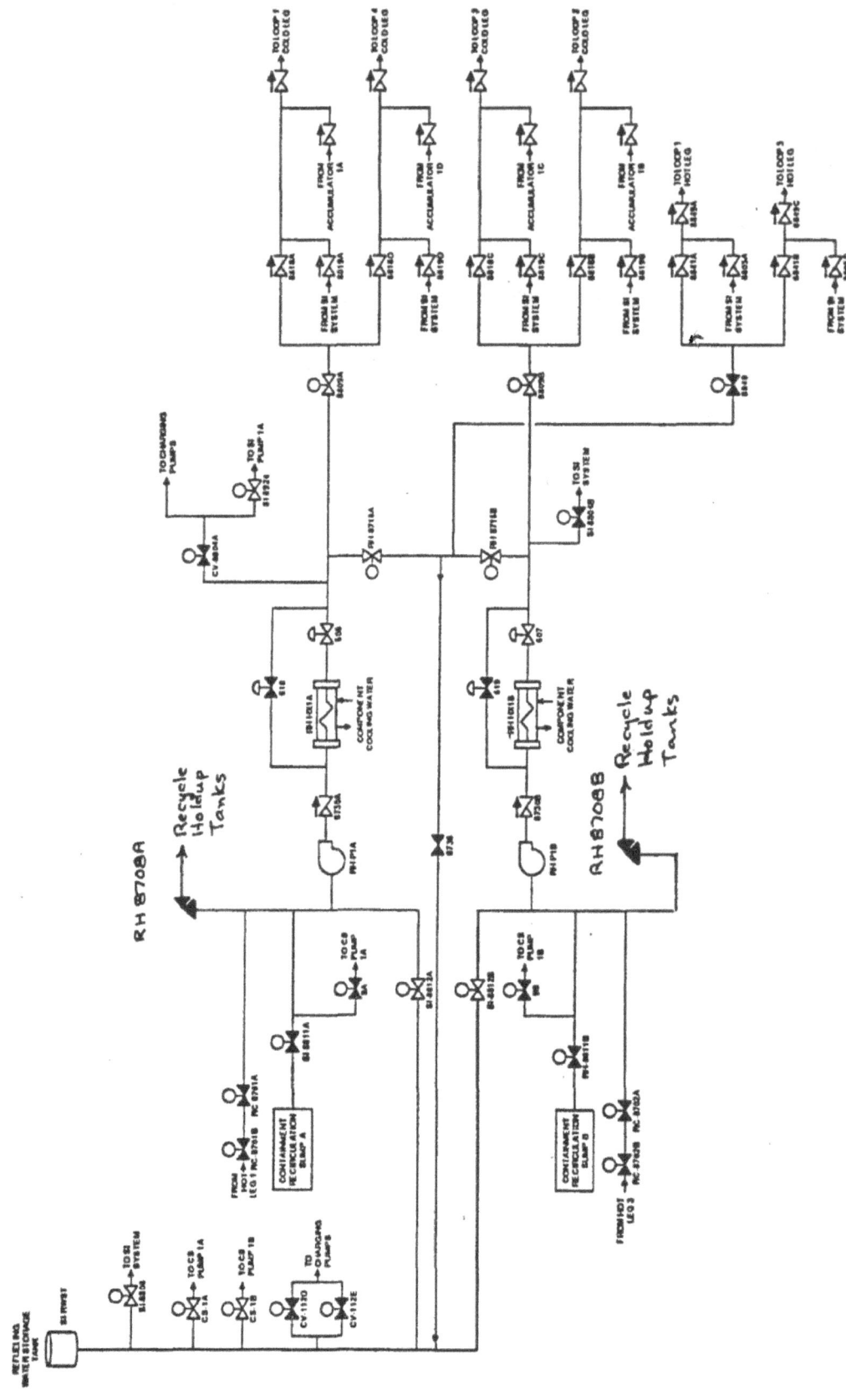

Fig. 1. Simplified drawing of the Braidwood RHR system

ATTACHMENT A

SEQUENCE OF EVENTS

(for LER 456/89-016)

DECEMBER 1, 1989

CENTRAL STANDARD TIME

*NOTE: The following sequence times are based on a collection of
the best information available during the inspection. Therefore,
there may be some variances with other information provided.*

0000   *Initial Conditions*: At the beginning of Shift 1, Unit 1 was in cold shutdown (Mode 5), RCS was solid with the temperature at 175F and pressure was 350 psig.

Operations personnel were in the process of drawing a bubble in the pressurizer. Reactor coolant pumps (RCPs) B and D were in operation with the pressurizer power operated relief valves in "cold over pressure protection" condition. 1A RHR pump (train) was in operation in the shutdown cooling mode with 1B RHR train idle and available for operation. The 1A charging pump was in normal operation with letdown coming from the RHR system. 1B RHR pump and both safety injection pumps were secured and tagged out of service as required by Technical Specifications and procedures for RCS cold over pressure protection. In addition, 1A RHR pump suction valve 1RHR 8701B was tagged out of service open with power removed by procedure to assure RHR would be maintained in the event of a pressure switch malfunction.

0055   Commenced drawing a bubble in the pressurizer by increasing letdown flow and energizing PZR heaters per BwOP RY-5, "Drawing a Bubble in the Pressurizer."

0128   RCS pressure had increased to about 395 psig. Letdown flow was increased to stabilize pressure.

0142   Letdown flow was maximized and charging flow was minimized (to about 70 gpm) to accommodate the RCS pressure increase to 404 psig as indicated on the wide range pressure instrument. Later it was found that the 1B RHR pump suction pressure had reached 416 psig. Although unknown at the time, this is where the 1B RHR pump suction relief valve is believed to have lifted.

0144   Pressurizer level reached on scale from off scale high and was decreasing rapidly. Letdown flow was reduced to stabilize pressurizer level.

0145   The radwaste operator informed the control room of a significant increase in holdup tanks (HUTs) levels.

0149   Charging flow was increased to correct for the rapid drop in pressurizer level.

Operations personnel manually swapped charging pump suction from the volume control tank (VCT) to the RWST.

0152   Pressurizer level went off-scale low.

0153   Charging flow was increased to maximum and letdown was reduced to minimum.

0155   1B RHR train cooling was started and 1A RHR train was secured and isolation started. This is based on field reports of a relief problem in the vicinity of the 1A RHR pump suction relief valve and accepted engineering practice to assume a fault is on the operating train.

0159   Secured 1B RCP due to primary pressure dropping to less than 325 psig and the lowest pump shaft seal differential pressure. 1D RCP continued to operate throughout the event. Primary system pressure was noted to be 272 psig and later verified by computer data to be the lowest RCS pressure throughout the event.

0215   1B charging pump out of service was lifted and was placed in operation to provide additional charging flow. This resulted in an associated RCS pressure increase.

0227   A GSEP "ALERT" was declared for loss of coolant inventory beyond the capability of the makeup system.

0235   1A RHR pump suction valve out of service was lifted and the valve shut to complete isolation of the IA RHR train and suspected leak.

0237   Nuclear Accident Report System (NARS) notification made to State of Illinois.

0245   Pressurizer level was identified as increasing on Channel LI462 and RCS pressure reached 310 psig.

        1B charging pump was secured. Radwaste reported HUT levels still increasing.

0254   Pressurizer level was identified as decreasing. 1B charging pump was restarted.

0302   Pressurizer level was increasing. Charging flow was reduced to slow the rate of pressurizer level increase and possible thermal shock to the pressurizer.

0319   An operator in the auxiliary building reported evidence of flow through the 1B RHR pump suction relief valve due to noise level and associated pipe temperatures (touch).

0322   Opened and closed 1RH 8734A (1A RHR cross connect to letdown) to reduce 1A RHR train pressure for assurance that the 1A RHR pump suction relief valve was shut.

0324   Resident Inspectors were notified.

0326   ENS notification to the NRC.

0335   Unit 1 shift foreman reported leakage, from the vicinity of relief valve OAB 8634 (discharge common to RHR pump suction reliefs to the HUTs). This was later determined to be from a weep hole in the side of the valve and was the source of the 30 to 50 gal of water released to a limited area of the auxiliary building.

0342   Charging flow was increased for adjustment to maintain pressurizer level.

0345   An operator was stationed near the 1A RHR pump suction relief valve.

0346   1A RHR train isolation valves were opened and locally verified that there was no evidence of flow through the 1A RHR pump suction relief valve.

0349   Placed the 1A RHR train in operation by starting the 1A RHR pump.

0350   Secured the 1B RHR pump and isolated the 1B RHR train.

0352   Pressurizer level showed significant increase.

0353   Secured the 1B CV pump.

0354   A field operator reported no evidence of leakage from the 1A RHR pump suction relief valve.

0356   A field operator reported no evidence of leakage from 1B RHR pump suction relief valve.

0400   Placed the 1A RHR letdown in service.

0402   Radwaste reported HUT levels had stabilized.

0415   Manually transferred charging pump suction from RWST back to the volume control tank.

0427   GSEP control transferred to Technical Support Center (TSC).

0435   GSEP "ALERT" terminated.

ACCIDENT SEQUENCE PRECURSOR PROGRAM COLD SHUTDOWN EVENT ANALYSIS

LER No.:	458/89-020
Event Description:	Freeze seal failure
Date of Event:	April 19, 1989
Plant:	River Bend

Summary

River Bend Station was in a refueling outage on April 19, 1989 when a freeze seal in the standby service water (SSW) system failed. When the seal was lost, water from the system was discharged from a disassembled 6" valve, and flowed across the floor and down to the next lower level in the building. A switchgear on the lower level was shorted out resulting in the loss of reactor protection system (RPS) Division II and subsequently the loss of a vital 120 V-AC power supply. The plant lost shutdown cooling (SDC) for 17 min, normal lighting for the reactor, control, and auxiliary buildings, a load center transformer, normal spent fuel pool cooling (SFPC) system, and a RPS motor generator (MG) set as a result of the 15,000 gal flood. Operators isolated the leak within 15 min. The conditional core damage probability estimated for this event is less than 1×10^{-6}.

Event Description

On April 19, 1989 work was being performed on the SSW supply (1SWP*V524) and return (1SW*V525) valves for unit cooler 1HVR*UC11B, since these valves were non-isolable, a freeze seal had been established so the valves could be disassembled. Two freeze plugs had been formed using one supply line from two liquid nitrogen sources. A freeze seal watch had begun, and 10 min after nitrogen supplies had been switched, a loud noise was heard by the person on watch. The supply line freeze plug had given way, but the return line plug remained in place and did so throughout the event. The control room was notified of leakage past a freeze seal. An operator sent to investigate the leak in the auxiliary building found water on the floor at the 114-ft elevation. He then proceeded to the 141-ft elevation and found water flowing across the floor and a 6-ft high column of water flowing from the body of the inlet isolation valve to cooler 1HVR*UC11B. The operator then assisted maintenance personnel trying to re-install the valve bonnet on the valve. This operator did not contact the control room to tell the operators of his assessment of the situation and the status of the leak. Water flowed from the 141-ft elevation to the 114-ft elevation through openings under motor control centers (MCCs) 2J and 2L. On the 114-ft elevation water entered load centers 1NJS-LDC 1A/B. The resulting ground faults in the load centers caused windings of the step-down transformer, 1NJS-X1A, to burn out and an electrical explosion in the adjacent 13.8 kV manual disconnect switch bay. Switchgear 1NPS-SWG1A Breaker 16 then opened and interrupted power to load centers 1NJS-LDC 1A, 1B, 1C, 1D, 1S, and 1T. This tripped RPS Bus "B" and resulted in a half scram and Division II containment isolation valves to close; thus, isolating SDC, tripping normal SFPC, tripping normal lighting to the reactor building, containment building, and auxiliary building. Operators then proceeded to restore SDC and SFPC using their abnormal operating procedures. Also, at this time, the shift supervisor (SS) and control operating foreman (COF) were trying to ascertain the source of the leak. After discussion and investigation, The SS and COF decided to isolate Division II of SSW and remove it from service. The SS and COF did this without positive confirmation that it was the leak source, but they had correctly inferred that it was the leak source from their investigation. Within minutes the leak stopped and the maintenance personnel re-installed the bonnet on the valve body that was leaking. Shortly thereafter, RHR SDC was restored using Division I RHR. Normal SFPC was restored about six h later.

The delay in restoring SFPC was due to re-establishing power to the component cooling water (CCW) pumps which were powered by the damaged 13.8 kV load center.

A drawing of the River Bend SSW system is provided in Fig. 1 and a drawing of Division I of RHR is provided in Fig. 2.

Additional Event-Related Information

Initial water level was 23 ft above the reactor vessel flange, this corresponds to about 640 in (or more than 53 ft) above top of active fuel (TAF). A reactor scram and automatic isolation of the RHR SDC from the reactor recirculation system occur at 172 in above TAF. Emergency core cooling system (ECCS) initiation occurs at 19 in above TAF. Upon ECCS initiation, RHR automatically lines up for and initiates in the low-pressure coolant injection (LPCI) mode. Also, both high-pressure core spray (HPCS) and low-pressure core spray (LPCS) systems automatically line up for and initiate vessel makeup.

Various pieces of equipment on the lower elevations of the auxiliary building were jeopardized by the flooding. As a result, the potential for flooding becoming a common mode failure mechanism through which redundant systems could be disabled was examined. The most limiting sequence of events was determined to be due to the inadequate capacity of the floor drains associated with the flooding of the lower elevations in the auxiliary building caused by the leak and/or from postulated fire fighting activities for electrical fires in transformers, switchgear, or MCCs resulting from the leak on higher elevations. If the drain system allowed the water to collect on the lower elevations, the safety-related equipment there would be jeopardized. However, it was determined that while three RHR/LPCI, the LPCS, and the HPCS pumps are all located on the lower elevations of the auxiliary building and it is possible following extensive unchecked flooding and/or fire fighting activities to put these pumps at risk, this was considered to be unlikely; moreover, only the LPCS and one RHR/LPCI pump were located directly below the leak. Flooding, in this case, posed little risk to the core.

ASP Modeling Assumptions and Approach

Analysis for this event was developed based on procedures (e.g. Procedure STP-204-0700, Rev. 1, effective March 3, 1989) in effect at River Bend at the time of the event, the Plant Technical Specifications, the Augmented Inspection Team (AIT) report, and the Final Safety Analysis Report (FSAR).

The following comments are applicable for the analysis of this event:

a. Core damage end state. Core damage is defined for the purpose of this analysis as reduction in reactor pressure vessel (RPV) level above TAF or failure to cool the suppression pool in the long term. With respect to RPV inventory, this definition may be conservative, since steam cooling may limit clad temperature increase in some situations. However, choice of TAF as the damage criterion allows the use of simplified calculations to estimate the time to an unacceptable end state.

b. Boil-off of RPV inventory can be delayed if RPV level can be raised to permit natural circulation. Availability of RPV injection to raise water level for natural circulation is included in the analysis.

c. Three pressure vessel head states were considered for the analysis: head on and tensioned, head on and detensioned, and head off. If the head is on and tensioned, then decay heat removal as well as vessel makeup methods which require pressurization are assumed to be viable. If the head is on, but detensioned, then failure to maintain the RPV depressurized is also assumed to proceed to core damage (this assumption is conservative). If the head is off, then makeup at a rate equal to boil-off is assumed to provide core cooling.

d. Five makeup sources were available during this event: HPCS, LPCI, LPCS, control rod drive (CRD) flow and the feedwater/condensate system. Use of any other source of makeup is considered to be a recovery action.

e. Successful termination of a loss of RHR (SDC) is defined as recovery of RHR or provision of alternate decay heat removal via the suppression pool or main condenser, or, if the head is removed, via refueling cavity boiling. Also, injection from any source to raise RPV level and allow natural circulation increases the amount of time available for recovery of RHR.

f. If RHR (SDC) cannot be recovered, then alternate core cooling methods are needed. If the head is tensioned, these involve allowing the RPV to repressurize, opening of at least one safety relief valve (SRV), and dumping decay heat to the suppression pool. If the condenser and condensate system are available, then decay heat can also be dumped to the condenser. If the head is detensioned, then decay heat must be removed without the RPV being pressurized. This requires opening of at least three SRVs and recirculating water to the suppression pool using the core spray or LPCI pumps. For all cooling modes involving the suppression pool, suppression pool cooling must be initiated in sufficient time to prevent the suppression pool from exceeding its temperature limit. If the head is removed, then any makeup source greater than ~200 gpm, combined with boiling in the RPV, will provide adequate core cooling.

The event tree model for this event is shown in Fig. 3. In the event, electrical faults from the flood resulted in RHR isolation. Isolation of Division II of SSW also rendered RHR Division II unavailable, since the two RHR heat exchangers in that division could not provide cooling. Because of these faults, the event has been modeled as a loss of SDC with one train of RHR (SDC) and suppression pool cooling unavailable. Note that these trains were recoverable once the bonnet on the open isolation valve was re-installed.

The event tree model includes the following branches:

Head Status. For the operational event in question, the head was off. However, since the event involved isolation of one auxiliary building cooler for valve maintenance with both SSW trains in operation, it was assumed that the event could have occurred with the head on as well. The likelihood of the three different head states was assumed to be:

head on	0.27
head detensioned	0.02
head off	0.71

These values are consistent with values developed for Washington Nuclear Plant, Unit 2, based on an analysis of shutdown outages for that plant.

LPCI or LPCS Flow Available. LPCI consists of three trains at River Bend. Each train includes one pump with a single normally-open suction valve and a single normally-closed discharge (RPV injection) valve. The pump's normal suction source is the suppression pool.

LPCS consists of one train at River Bend. This train includes one pump with a single, normally open motor-operated suction valve and a single normally-closed discharge (RPV injection) valve. The pump suction source is normally the suppression pool.

To simplify the estimation of the probability of failure of suppression pool cooling (which is dependant on the LPCI trains which also provide RHR), only the probability of failure of core spray and the probability of failure of the "C" train of LPCI was used to estimate this branch probability. Assuming that neither the LPCS nor LPCI pumps require SSW for injection, and that normally-open valves and check valves do not contribute substantially to system unavailability, the equation for this event tree branch is therefore

(LPCS-P1 + LPCS-5) * (LPCI-P2C + LPCI-42C).

Applying screening probabilities of 0.01 for failure of a motor-driven pump to start-and-run and failure of a motor-operated valve to open, 0.1 for the conditional probability of the second similar component to operate, and a probability of not recovering the faulted branch, results in an overall failure probability for the branch of 7.5×10^{-4}.

HPCS Flow Available. HPCS consists of one train at River Bend. This train includes one pump with a single, normally-open motor-operated suction valve and a single normally-closed discharge (RPV injection) valve. The pump suction is normally the condensate storage tank. Making the same assumptions as for the previous branch results in a failure probability estimate of 6.8×10^{-3}.

CRD Flow Available. At cold shutdown pressures, one of two CRD pumps can provide makeup. Since one pump is typically running, the system will fail if that pump fails to run or if the other (standby) pump fails to start and run. Assuming a probability of 0.01 for failure of the standby CRD pump to start, and 3.0×10^{-5}/hr for failure of a pump to run, results in an estimated failure probability for CRD flow of 2.5×10^{-6}. In this estimate, a short-term non-recovery likelihood of 0.34 was applied to the non-running pump failure-to-start probability, consistent with the approach used to estimate the failure probability for the core spray system. A mission time of 24 h was also assumed.

If only one train is available (because of maintenance on the opposite division), then the CRD failure probability is estimated to be 7.2×10^{-4}.

Feedwater/Condensate Available. River Bend has three motor-driven feedwater and three motor-driven condensate pumps; and, while the condensate pumps can provide more than adequate makeup, they are often unavailable during a refueling outage because of work on the secondary system. However, for this event, the feedwater/condensate system was available. A failure probability of 0.01 was assumed.

RHR (SDC) Recovered (Short Term). For River Bend, RHR is available provided RPV level is greater than the low-level isolation level and RPV pressure is less than the high-pressure isolation pressure. For events with the head on and for which reactor vessel inventory was increased to provide for natural circulation, RHR must be recovered, if lost, prior to reactor vessel pressure reaching the high-pressure isolation setpoint (135 psig at River Bend), which would prevent opening the suction line isolation valves and restoring RHR. Once the high-pressure isolation setpoint is reached, operation of at least one SRV was assumed to be required, and the event proceeds with RPV depressurization and the use of RHR in the suppression pool cooling mode to remove decay heat. The main concern, then, is the time from the initial loss of RHR until the high-pressure isolation setpoint is reached, and for events with the head on but with short-term makeup unavailable, this time period is even more restrictive.

If the head is detensioned, the time period to recover RHR is assumed to be the time to reach boiling, and usually this is the most limiting time period.

If the RPV head is off, as was the case for this event, it is estimated based on simplifying assumptions that the water above the core would not reach boiling for approximately four d, and it would be more than 25 d before the core would be uncovered. This very long time is attributable to the enormous vessel inventory available above TAF (23 ft above the flange), the equally large volume of water available from the spent fuel pool, and the relatively low decay heat load from the core 36 d after shutdown.

During this event, SDC was recovered by transferring RPS bus "B" to its alternate supply, which allowed the Division II containment isolation signal to be reset and the SDC isolation valves to be opened. This action was performed in 17 min. Considering the time period available for SDC recovery, ample time exists to accomplish this action. Therefore, the probability of failing to recover SDC was estimated based only on component failure likelihoods, without consideration of any associated human errors.

Since one of the two RHR trains was unavailable because of the isolation of its associated SSW train, both suction isolation valves must open and the remaining-train RHR pump must start and run for RHR success. Using the same screening probabilities as for the earlier branches, a failure probability of 1.0×10^{-2} is estimated.

Main Condenser Available. The main condenser is modeled as a heat removal mechanism for sequences in which the condensate system is used as an injection source and the head is tensioned. The probability of the condenser being available for heat removal, given the condensate system is available, was assumed to be 0.5. The actual likelihood is dependant on the nature of the outage.

Required SRVs Opened. Sixteen SRVs [seven of which are also designated automatic depressurization system (ADS) valves] are installed at River Bend. For events with the head tensioned, opening of one or more SRVs is assumed to provide success in mitigating a loss of RHR (SDC). For events with the head detensioned but still on the vessel opening of three SRVs are assumed to be required for success. The number of valves which are assumed to be required is based on calculations done at Pennsylvania Power and Light for Susquehanna. In either case, failure of the valves to operate is dominated by dependant failure effects.

A probability of 1.6×10^{-4} was used for failure of multiple SRVs to open. This value was based on the observation of no such failures in the 1984-1990 time period, combined with a non-recovery likelihood of 0.12. This approach is consistent with the approach used to estimate this probability for other ASP evaluations, but includes a longer observation period and a lower probability of failing to recover to account for the 4-6 h typically available to open the valves [a non-recovery value of 0.71 is used for the probability of not recovering an ADS actuation failure in a one-half hour time period (see NUREG/CR-4674, Vol. 6) — this value was also used to estimate the likelihood of SRV failure for sequences with the head detensioned but on, since time periods for these sequences are short].

A value of 1.6×10^{-4} is consistent with failure probabilities which can be estimated from individual valve failure probabilities and beta factors, as described in NUREG/CR-4550, Vol 1, Rev. 1, "Analysis of Core Damage Frequency: Internal Events Methodology," and the conditional probability screening values used in the ASP program. The failure probabilities estimated using either approach are probably conservative, considering the number of valves potentially available for use. (NUREG/CR-4550, Vol 4, Rev. 1, Part 1, "Analysis of Core Damage Frequency: Peach Bottom, Unit 2, Internal Events," used a value of 1.0×10^{-6} for common cause SRV hardware faults, based on engineering judgement.

Suppression Pool Cooling (Long-Term). Suppression pool cooling at River Bend, like most BWRs, is a mode of RHR. RHR consists of three independent loops at River Bend, and each loop contains its own motor-driven pump, has a suction from the suppression pool, and is capable of discharging water to the reactor vessel via a separate nozzle or back to the suppression pool via a full-flow test line. Two of these loops have two heat exchangers which are cooled by normal or standby service water. For these two loops, one or more RHR/LPCI pumps take suction from the suppression pool, pump water through the heat exchangers if necessary, and return it to the suppression pool. The suppression pool cooling mode of RHR consists of two redundant trains, each of which includes an RHR/LPCI pump, two series heat exchangers, and a single return valve which must be opened to return flow to the suppression pool. For the train providing RHR (SDC), the suppression pool suction valve [normally open for LPCI but closed for RHR (SDC)] must also be opened to provide suction to its respective pump. During this event, RHR loop B was providing shutdown cooling, and hence opening of suction valve E12*MOVF004B was assumed to be required for this mode of operation.

Since one of the two RHR trains was initially unavailable (because of the isolation of its SSW train), the RHR pump in the remaining train must start and run, its suppression pool suction valve must open, and its discharge valve (E12*MOVF024B) to the suppression pool must open. In addition, one of the suction valves from the reactor recirculation loop and one of the normal RHR injection valves must close. If this train fails to provide suppression pool cooling, then the initially faulted train must be recovered. A branch probability of 0.03 is estimated, conditional on the failure to recover RHR (SDC) in the short term.

It should be noted that, because of the length of time available to recover suppression pool cooling (greater than 24 h), and the general lack of understanding of the reliability of such actions, this time estimate has a high degree of uncertainty associated with it.

Analysis Results

Branch probabilities developed above were applied to the event tree model shown in Fig. 3 to estimate a conditional probability of subsequent severe core damage for the event at River Bend. This conditional probability is much less than 1.0×10^{-6}, based on the head state (removed) which existed during the event. Branch probabilities are shown on Fig. 3. The dominant sequence involves failure to provide RPV makeup from one of the variety of sources in the long term.

An additional calculation was performed to determine the impact of head status on the conditional probability estimate. If the event could have occurred with the head on, detensioned but on, or off (with probabilities as previously specified), then the conditional probability for the event is estimated to be much higher, $\sim 9.0 \times 10^{-5}$. This high probability is a result of the two train design of the RHR system on this plant, and the component failure probabilities assumed in the analysis.

Flooding in the auxiliary building was examined and it was determined that the RHR/LPCI and LPCS systems would only suffer a loss of redundancy if the flooding were allowed to proceed unchecked. Since it was unlikely that extensive flooding would have occurred during this event, this analysis did not perform a complete flooding analysis. Even if a hypothetical flood, such as the one posed by the AIT investigation, of the auxiliary building had occurred which failed all the ECCS equipment located on the lower elevation, both the CRD and condensate systems were available for vessel makeup. Several things happened during this event that would have mitigated extensive flooding. First, no electrical fire occurred, so flooding from fire fighting activities was not possible. Second, maintenance personnel in the area of the failed freeze seal were in the process of reassembling the valve when the control room operators remotely isolated the leak, and these maintenance technicians would have been able to stop the leak within minutes if no remote isolation had occurred. Third, the flooding that did occur only impacted a single division of ECCS. Lastly, the leak was confined mostly to the upper elevations since there was only one small flow path to the lower elevations. Therefore, it is unlikely that other ECCS equipment would have been jeopardized.

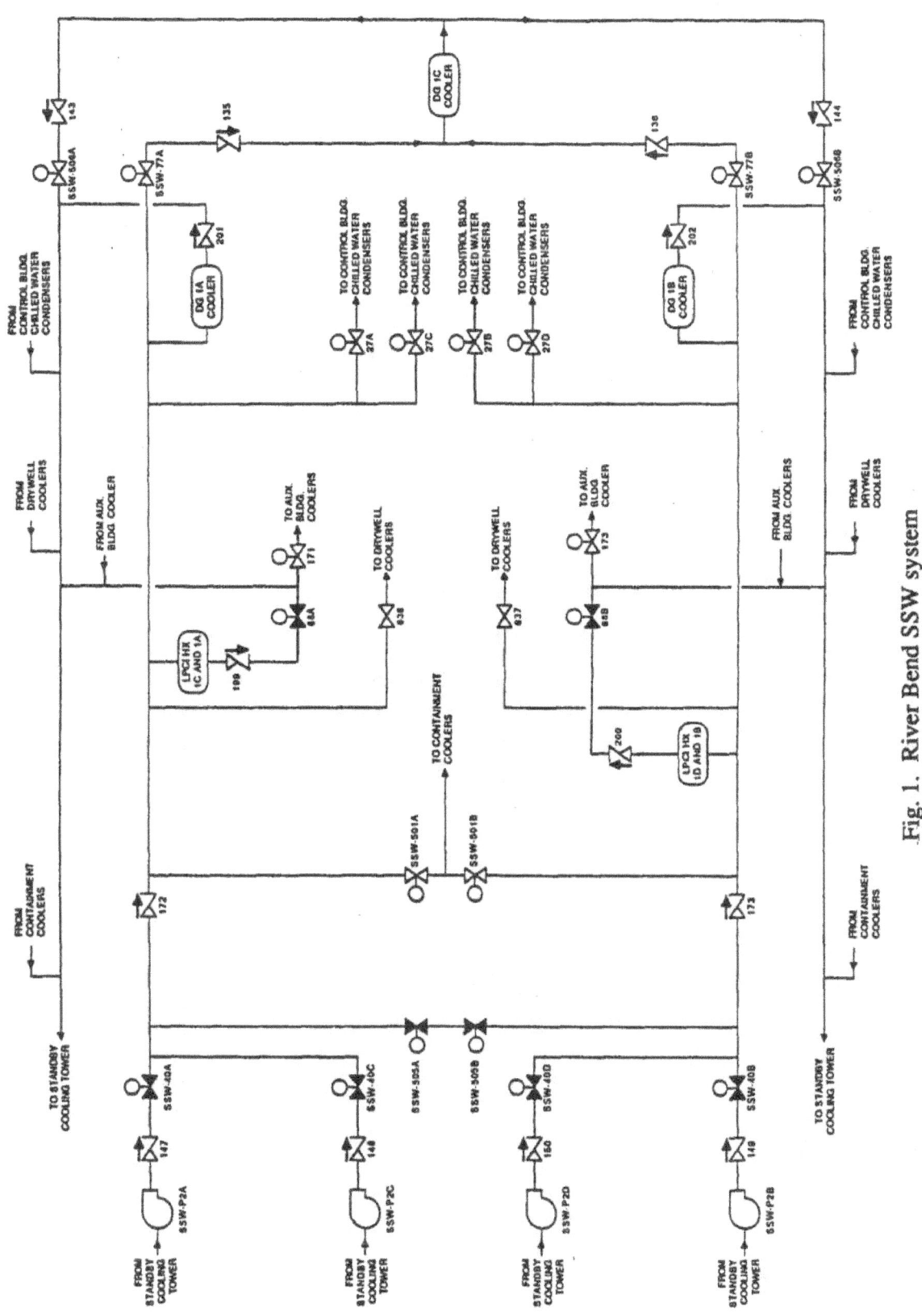

Fig. 1. River Bend SSW system

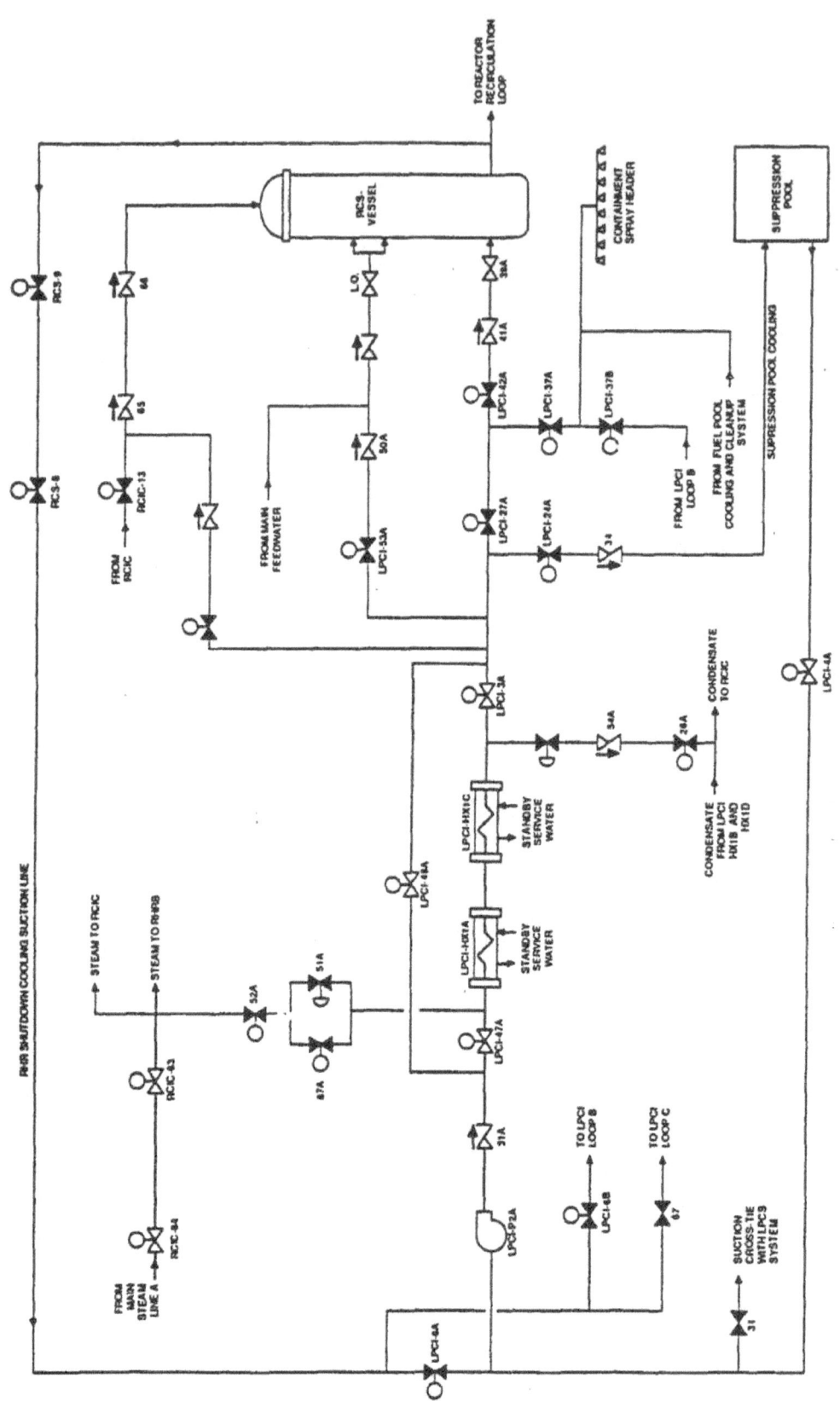

Fig. 2. River Bend RHR system loop A

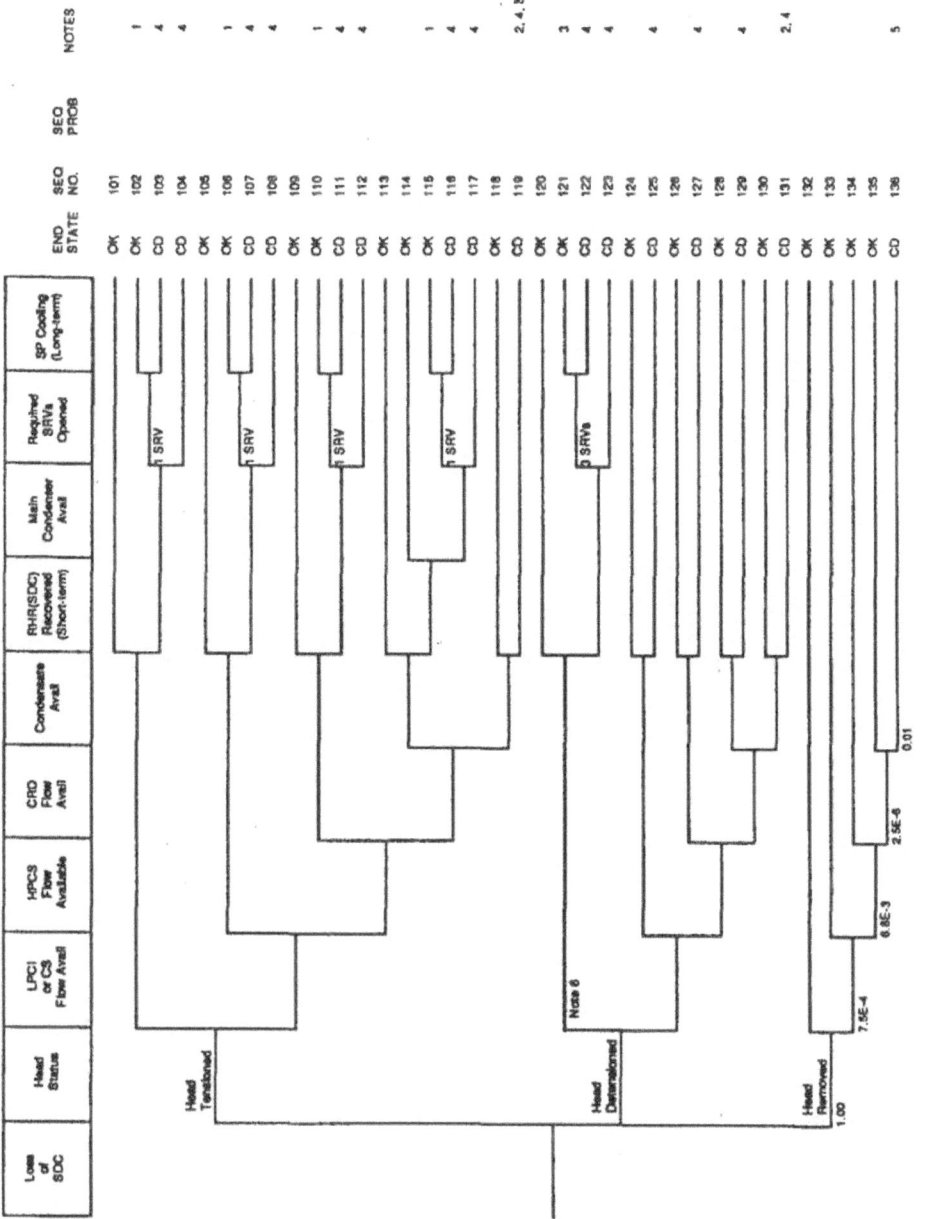

Figure 3. Event tree model for LER 458/89-020

Notes: 1. Suppression pool level will increase in this sequence.
2. Reduced time to recover RHR if recirculation pump unavailable since makeup required to achieve natural circulation is also unavailable.
3. Water in main steam lines may overstress these lines.
4. Use of RWCU/Condensate Transfer to transfer hot water to the condenser or condensate storage tank will increase the time available to recover RHR or initiate suppression pool cooling.
5. Alternate injection sources such as service water may also provide injection.
6. If primary and secondary containment cannot be established, this sequence is prescribed.

APPENDIX B

Details of Equipment Hatch Survey

Table B.1 Boiling-Water Reactors

Plant & (OL date)	Containment type	Hatch type[1]	No. of bolts	Additional inspection for refueling closure[2]	Temporary platform[3]	Air or ac needed	Bolt pattern[4]	Comments
Big Rock Pt. (64)	Sphere	In	N/A	App. J. Type B	No	ac	N/A	TS requires containment when fuel is in reactor.
Browns Ferry (73/74/76)	Mark I	In[5]	12	No	Ladder	Manual	Holddown clamp	Double door
Brunswick 1&2 (76/74)	Mark I	In	12	No	No	Manual	B	
Clinton (87)	Mark III	In	20	No	Yes	Manual	B	
Cooper (74)	Mark I	In	8	No	No	None	A	
Dresden 2&3 (69/71)	Mark I	In	8	No	No	Manual	B	
Duane Arnold (74)	Mark I	In	12	No	Yes	ac	B	Need ac for crane to install hatch
Fermi (85)	Mark I	Out/In	20/36	No	Yes	Manual	B	Two equipment hatches
FitzPatrick (74)	Mark I	In	8	No	No	Manual	B	
Grand Gulf (84)	Mark III	In	20	No	No	ac	B	
Hatch 1&2 (74/78)	Mark I	In	8	No	Yes	Manual	B	Can close hatch without temporary platforms.
Hope Creek (86)	Mark I	In	24	No	Yes	Manual	B	
LaSalle 1&2 (82/84)	Mark II	In	16	No	No	Manual	B	
Limerick 1&2 (85/89)	Mark II	Out	80	No	Yes	ac	B	
Millstone 1 (86)	Mark I	In	8	No	Ladder	Manual	B	

See footnotes at end of table.

Table B.1 (cont.)

Plant & (OL date)	Containment type	Hatch type[1]	No. of bolts	Additional inspection for refueling closure[2]	Temporary platform[3]	Air or ac needed	Bolt pattern[4]	Comments
Monticello (81)	Mark I	In	8	No	No	Manual	B	
Nine Mile Pt. 1 (74)	Mark I	Out	36	No	Yes	Manual	B	Inspector noticed a gap with minimum bolts installed.
Nine Mile Pt. 2 (87)	Mark II	Out	64	No	Yes	Manual	B	
Oyster Creek (69)	Mark I	In	36	No	No	Air	B	
Peach Bottom 2&3 (73/74)	Mark I	In	8	No	No	Manual	B	
Perry (86)	Mark III	Out	72	No	Yes	ac	A	
Pilgrim (72)	Mark I	Out	8	No	No	No	A	Licensee noted speedy closing difficult due to temporary services.
Quad Cities1&2 (72/72)	Mark I	In	8	No	Yes	Manual	B	
River Bend (85)	Mark III	Out	64	No	No	Manual	A	
Susquehanna 1&2 (82/84)	Mark II	Out	30	No	No	Air & ac	B	Can close hatch manually.
Vermont Yankee (73)	Mark I	Out	8	No	No	Manual	B	
WNP–2 (84)	Mark II	Out	64	No	No	Air	A	Can close hatch manually.

[1]Hatch type: Out = pressure-unseating design; In = pressure-seating design.
[2]A confirmatory inspection done voluntarily by some licensees to verify that the hatch is seated properly.
[3]Temporary platforms are used in some plants for workmen to reach the bolts.
[4]Bolt pattern: A = bolt in threaded hole; B = bolt swing.
[5]Flat plate.

Table B.2 Pressurized-Water Reactors

Plant, [Vendor], & (OL date)	Containment type	Hatch type[1]	No. of bolts	Additional inspection for refueling closure[2]	Temporary platform[3]	Air or ac needed[4]	Bolt pattern[5]	Comments
Arkansas 1 [B&W] (74)	Large dry	In	4/24	None	No	Manual	B	
Arkansas 2 [CE] (78)	Large dry	In	4/16	None	No	Manual	B	No procedure for temporary closing; just tighten bolt, close opening.
Beaver Valley 1&2 [W] (76/87)	Subatmospheric	In	4/24	None	Ladder	Manual	B	Emergency airlock inside hatch.
Braidwood 1&2 [W] (87/88)	Large dry	In	0/20[6]	None	Yes	ac	B	Have loop ISO vales, don't drain to midloop.
Byron 1&2 [W] (85/87)	Large dry	In	0/20[6]	None	Yes	ac	B	
Callaway [W] (84)	Large dry	In	4/20	None	No	ac	B	Special rigging needed to close hatch during station blackout.
Calvert Cliffs 1&2 [W] (74/76)	Large dry	In	4/20	None	No	ac	B	
Catawba 1&2 [W] (85/86)	Ice condenser	In	4/16 4/24	None	No	ac	B	Unit 2 modified to add bolts to seal.
								Inspector notes increased number of bolts used for fuel move to close gap. Unit 1 uses 10, Unit 2 uses 15 bolts.
Comanche Peak [W] (90)	Large dry	In	4/16	None	Ladder	Manual	B	
Cook 1&2 [W] (74/77)	Ice condenser	Out	0/32[6]	None	No	ac	A	No requirement for hatch but licensee maintains it for fuel move & midloop.
Crystal River [B&W] (77)	Large dry	Out	4/72	None	Yes	Air	B	Hatch can be closed manually with truck-mounted crane.
Davis-Besse [B&W] (77)	Large dry	In	4/12	None	Yes	Manual	B	

See footnotes at end of table.

Table B.2 (cont.)

Plant, [Vendor], & (OL date)	Containment type	Hatch type[1]	No. of bolts	Additional inspection for refueling closure[2]	Temporary platform[3]	Air or ac needed[4]	Bolt pattern[5]	Comments
Diablo Canyon 1&2 [W] (84/85)	Large dry	In	4/48	Daylight check	Ladder	Manual	B	Perform daylight check. One seal may be used for Modes 5 & 6.
Farley 1&2 [W] (77/81)	Large dry	In	4/28	None	Yes	Manual	B	
Fort Calhoun [CE] (73)	Large dry	In	4/36	None	No	ac	B	
Ginna [W] (84)	Large dry	Out	36/36	QC metal	Yes	Manual	B	Licensee uses a temporary closure plate for temporary services.
Haddam Neck [W] (74)	Large dry	Out	18/92	None	No	ac	B	Mobile crane can be used to install hatch.
Harris [W] (87)	Large dry	Out	4/36	None	Ladder	Manual	A	
Indian Pt. 2 [W] (73)	Large dry	In	20/20	None	No	ac[7]	B	Licensee has a temporary closure plate for temporary services.
Indian Pt. 3 [W] (76)	Large dry	In	20/20	None	No	ac[7]	B	Licensee has no temporary closure plate.
Kewaunee [W] (73)	Large dry	In	12/12	None	-[8]	ac	A	Use boatswain chair to close hatch.
Main Yankee [CE] (73)	Sphere	Out	8/74	None	No	Manual	A	Mobile crane used to install hatch.
McGuire 1&2 [W] (81/83)	Ice condenser	In	4/16	None	Ladder	Manual	Holddown clamp	Noticed gap with 4 & 8 bolts in place.
Millstone 2 [CE] (86)	Large dry	In	4/20	None	Yes	Manual	B	
Millstone 3 [CE] (86)	Subatmospheric	In	4/16	None	No	Manual	B	
North Anna 1&2 [W] (78/80)	Subatmospheric	In	4/20	None	No	Manual	B	Licensee requires every 2nd bolt be installed.

See footnotes at end of table.

Plant	Containment	Inside/Outside	Bolts	Test	Ladder/Access	Closure	Class	Comments
Oconee 1, 2, & 3 [B&W] (73/73/74)	Large dry	In	4/48	None	No	ac	B	Can position hatch without power.
Palisades [CE] (72)	Large dry	In	0/24[8]	None	Ladder	Manual	B	Procedures to discontinue temporary services on loss of shutdown cooling.
Palo Verde 1, 2, & 3 [CE] (85/86/87)	Large dry	In	4/32	Ran integrated leak rate test with 8 bolts	No	ac	B	Can close hatch manually. Ran integrated leak rate test with 8 bolts.
Point Beach 1&2 [W] (70/73)	Large dry	In	66/66	None	No	Manual	B	Licensee closes hatch on reduced inventory.
Prairie Island 1&2 [W] (74/74)	Large dry	In	0/12[6]	App. J. Type B	Ladder	Manual	B	TS does not specify number of bolts
Robinson [W] (70)	Large dry	Out	8/48	None	Ladder	Manual & mobile crane	B	Ladders are secured near hatch. 80-ton mobile crane used for closing hatch.
Salem 1&2 [W] (76/81)	Large dry	In	4/16	None	Yes	ac	B	Has a hatch seal penetration system.
San Onofre 1 [W] (67)	Sphere	In	0/12	None	No	Manual	B	Licensee & inspector noticed gap with 4 bolts installed. Unit 1 refuels through hatch.
San Onofre 2&3 [CE] (82/83)	Large dry	In	4/16	None	No	ac	B	Close hatch quickly on station blackout.
Seabrook [W] (90)	Large dry	In	4/32	None	Yes	ac crane	B	4 hr to close hatch on station blackout.
Sequoyah 1&2 [W] (80/81)	Ice condenser	In	4/20	None	No	ac winch	B	Recently completed 1st refueling.
South Texas 1&2 [W] (88/89)	Large dry	In	4/28	None	No	ac	B	Can use chain fall in place of winch.
St. Lucie 1&2 [CE] (76/83)	Large dry	Out	4/12	None	No	ac	B	

Table B.2 (cont.)

Plant, [Vendor], & (OL date)	Containment type	Hatch type[1]	No. of bolts	Additional inspection for refueling closure[2]	Temporary platform[3]	Air or ac needed[4]	Bolt pattern[5]	Comments
Summer [W] (82)	Large dry	In	4/30	App J Type B	Ladder	ac	B	Integrated leak rate test with 4 bolts. Can close hatch without ac power.
Surry 1&2 [W] (72/73)	Subatmospheric	In	4/36	None	No	Manual	B	Licensee has temporary cover plate used for auxiliary services.
TMI-1 [B&W] (74)	Large dry	Out	4/72	None	Yes	Manual	B	Emergency hatch common with equipment hatch and mounted on carriage.
Trojan [W] (75)	Large dry	In	4/20	None	No	No	B	Procedure to close hatch during station blackout.
Turkey Pt. 3&4 [W] (72/73)	Large dry	In	4/58	None	No	Air	A	Hatch can be positioned manually.
Vogtle 1&2 [W] (87/88)	Large dry	In	4/30	None	No	ac	B	Can close hatch during station blackout.
Waterford [CE] (85)	Large dry	In	4/16	None	Yes	Manual	B	
Wolf Creek [W] (85)	Large dry	In	4/20	None	No	ac	B	
Yankee Rowe [W] (63)	Sphere	In	4/56	None	No	ac	B	
Zion 1&2 [W] (73/73)	Large dry	In	0/12[6]	Seal press. system	No	ac/air	B	Licensee can install hatch in 2 hours during station blackout. Hatch installed during midloop.

[1] Hatch type: Out = pressure-unseating design; In = pressure-seating design.
[2] A confirmatory inspection done voluntarily by some licensees to verify that the hatch is seated properly.
[3] Temporary platforms are used in some plants for workmen to reach the bolts.
[4] If neither ac power or air is required, the equipment hatch is closed manually.
[5] Bolt pattern: A = bolt in threaded hole; B = bolt swing.
[6] Zero bolts required during refueling because hatch opens to fuel handling building.
[7] Polar crane.
[8] Crane and boatswain chair.

APPENDIX C

Staff Responses to Comments Received
on Draft NUREG–1449

Table C.1 Staff Responses to Comments Received on Draft NUREG-1449

No.	Subject	Organization	Comment	Response
1	Clarification	BWROG	Define the term "integral RCS" in Section 7.2(4) of NUREG-1449.	The term "integral RCS" describes the reactor coolant system when the reactor coolant pressure boundary is intact.
2	Clarification	BWROG	Clarify statement on page 7-6 regarding containment closure plans for BWRs.	The statement reflects observations during visits to BWR plants that those plants do not have contingency plans for closing the primary containment in an emergency.
3	Clarification	BWROG	Comments at top of page 6-13 regarding ECCS operability not consistent with BWR/4 STS 3.5.2.	Statement on page 6-13 has been modified to correctly reflect the requirements in Section 3.5.2 of the current BWR/4 and BWR/6 STS.
4	Clarification	BWROG	Clarify statements on page 6-12 regarding methods of RHR in BWRs.	Additions and corrections have been made in Section 6.7 to reflect the wide variability among BWRs regarding methods of RHR.
5	Clarification	BWROG	Several statements in Chapter 5 on TS incorrect. Need clarification.	Appropriate corrections have been made in Section 5.1 to properly reflect current BWR/4 STS in the areas of reactivity control, low-pressure ECC subsystems, and containment atmosphere.
6	Clarification	BWROG	Clarify Section 7.2(4)(B) in NUREG-1449, i.e., proposed relaxation of TS action to go to cold shutdown.	This refers to the language in some PWR technical specifications that requires entry into MODE 5 (<200 °F) (<93 °C) from MODE 4 when an RHR train is declared inoperable and RCS loops are still available for heat removal via forced or natural circulation.
7	Clarification	BWROG	Add statement to Chapter 5 which references the BWR STS used for the evaluation.	General statements in Chapter 5 regarding BWR technical specifications are based on current BWR/4 standard technical specifications. A statement to this effect has been added to Section 5.1.1.
8	Clarification	FP&L	Clarify the Turkey Point event listed in Table 2.6. The boric acid flow path to charging pumps was available.	Statement was added to the NUREG indicating that boric acid flow path from the refueling water storage tank to the reactor via the charging pumps was available during the event.
9	Clarification	GEORGIA POWER	Section 6.12.2 needs clarification. It may be at odds with plant-specific emergency plans.	Section 6.12.2 in NUREG-1449 has been modified to clarify that the 30-minute time period only applies to the accountability of site personnel and not to the evacuation of nonessential personnel.
10	Correction	IND/MICH POWER	Correct information in Appendix B regarding the equipment hatch at D.C. Cook.	Corrections were made in Appendix B of NUREG-1449.

Table C.1 (cont.)

No.	Subject	Organization	Comment	Response
11	Clarification	NORTHEAST UTIL	Clarify statement on page 2–10 regarding ASP analysis.	Statement was revised for greater clarity.
12	Correction	NORTHEAST UTIL	Delete last paragraph of Section 2.2.2 on page 2–11.	The staff agrees with the statement and it has not been deleted. The staff and its contractor wish to discourage comparison of conditional core melt frequencies for at-power and shutdown events modeled in the ASP program.
13	Correction	NORTHEAST UTIL	On page 6–12, in the first paragraph, the term "reactor protection system" is used incorrectly.	The term "reactor protection system" has been replaced with the terms "primary containment and reactor vessel isolation system."
14	Clarification	NORTHEAST UTIL	Regarding discussion on page 6–7; adequate defense in depth can be obtained by other than containment integrity.	The staff agrees and page 6–7 has been modified. During nonaccident conditions, a passive method of subcooled decay heat removal can suffice while normal decay heat removal systems are being restored and precludes the need for containment integrity.
15	Clarification	NUMARC	Differentiate between primary and secondary containment in NUREG–1449.	The report has been revised to clarify containment function.
16	Correction	TVA	Correct the title of Table 2.3 and Table 2.4.	Titles for Tables 2.3 and 2.4 have been corrected.
17	Fire Protection	BWROG	The benefit of a shutdown fire hazards analysis is not demonstrated given current requirements and NUMARC 91–06.	The staff disagrees. NUMARC 91–06 does not address fire protection.
18	Fire Protection	DETROIT-EDISON	NUREG–1449 does not take into account that TS require fire protection equipment to be OPERABLE when equipment protected is OPERABLE.	The staff disagrees. Equipment required to be operable was considered.
19	Fire Protection	ENTERGY OP. INC	The need for a fire hazards analysis is not demonstrated. A better approach is to have strong administrative controls.	The staff disagrees. A focused analysis based on realistic assumptions is appropriate.
20	Fire Protection	NORTHEAST UTIL	Shutdown fire hazards analysis is not needed. Existing fire hazards analysis, better TS, and NUMARC 91–06 are sufficient.	The staff disagrees. See responses to comments 17 and 19.
21	Fire Protection	NUMARC	Fire hazards analysis for shutdown is not necessary. Hazards will be considered in implementation of NUMARC 91–06.	The staff disagrees. See responses to comments 17 and 19.

#	Category	Organization	Comment	Staff Response
22	Fire Protection	NUMARC	We disagree with the observation that there are fewer fire protection controls during shutdown.	Observation is based on sample inspections. Staff agrees this may not be the case at all sites.
23	Fire Protection	NUMARC	Fire risk is not greater during shutdown; staff should reevaluate the basis for its conclusions.	The staff disagrees. As noted in Section 6.10, increased fire hazard conditions have been observed which would increase the probability that fire development and consequences would have increased significance.
24	Fire Protection	TVA	A fire hazards analysis and improved controls are not necessary because fires are not significant contributors to shutdown events.	The staff disagrees. See response to comment 19.
25	Fire Protection	WESTING-HOUSE	The NRC staff should assess the severity of past fires during shutdown before developing requirements.	The staff agrees that the industry has not suffered a serious fire during shutdown. However, precursor fires (e.g., Brunswick and Browns Ferry) have occurred.
26	Fire Protection	WESTING-HOUSE	We disagree with the statement that "..(fire) risk during shutdown is greater than for power operations."	The staff disagrees. See response to comment 23.
27	Fire Protection	YANKEE ATOMIC	The staff's statement that fire protection requirements do not apply to shutdown is incorrect.	The staff disagrees. Protection during power operation is the intent of the regulation. Exemption of the RHR system from protection presumes that hot shutdown can be reached and maintained. This may not be possible from a cold shutdown condition.
28	Fire Protection	YANKEE ATOMIC	The staff's statement that the probability of serious fire is greater during shutdown is not supported.	The staff disagrees. See response to comment 23.
29	Regulatory Action	ENTERGY OP. INC	NRC should provide additional guidance on meeting GL 88-17 given statements in NUREG-1449.	Additional guidance will not be provided. Individual issues will be resolved through the inspection program.
30	Regulatory Action	ENTERGY OP. INC	The staff should work with industry to develop a performance indicator. Design differences are important.	The staff agrees. It is NRC policy to solicit input from industry and the general public when developing performance indicators.
31	Regulatory Action	ENTERGY OP. INC	Regulatory changes should be reviewed against forthcoming NRC PRA results prior to promulgation.	The staff disagrees. This approach is not receiving serious consideration due to the uncertainty in the PRA models, the age of the database, and the narrow scope (i.e., 2 plants that are less than typical).
32	Regulatory Action	NORTHEAST UTIL	We agree with staff's conclusion not to address shutdown in the IPE program.	N/A

Table C.1 (cont.)

No.	Subject	Organization	Comment	Response
33	Regulatory Action	NORTHEAST UTIL	The NRC should not take regulatory action until the effectiveness of NUMARC 91-06 is evaluated.	The staff is considering this approach in its current regulatory analysis.
34	Regulatory Action	NUMARC	NRC should assess effectiveness of industry actions prior to proposing requirements in some areas.	The staff is considering this approach in its current regulatory analysis.
35	Regulatory Action	NUMARC	NRC should assess industry actions and regulatory requirements collectively, not as separate items.	The staff is doing this as part of its current regulatory analysis.
36	Regulatory Action	NUMARC	NUREG-1449 does not recognize that it will take time for industry actions to take effect.	The staff recognizes that it will take time for industry actions to take effect. The staff is considering this in its current regulatory analysis.
37	Regulatory Action	PACIFIC G&E	Rulemaking would overlap and conflict with industry initiatives. Allow industry time to make improvements.	The staff is considering the potential for conflict between potential regulatory requirements and industry initiatives in its current regulatory analysis.
38	Regulatory Action	PHIL ELECTRIC	Assess the effectiveness of industry action (by monitoring precursor events) prior to taking regulatory action.	The staff is considering this approach in its current regulatory analysis.
39	Regulatory Action	TVA	Issuing new requirements could decrease the effectiveness of industry efforts to address shutdown.	See response to comment 37.
40	Regulatory Action	YANKEE ATOMIC	NUREG-1449 studies do not support the need for new requirements.	The staff is currently performing a formal regulatory analysis to determine the need for new requirements. Studies documented in NUREG-1449 are being considered with other inputs in this analysis.
41	Regulatory Action	YANKEE ATOMIC	Refrain from regulatory action until industry's initiatives are implemented and assessed.	The staff is considering this approach in its current regulatory analysis.
42	Regulatory Action	YANKEE ATOMIC	Licensees should evaluate NUREG-1449 and factor information into the outage planning process.	The staff agrees.
43	Regulatory Action	YANKEE ATOMIC	NRC currently has regulatory authority to deal with poor performance during shutdown operations.	For some circumstances, and to a limited degree, this is true. However, backfitting in accordance with 10 CFR 50.109 would be necessary to impose clearly enforceable requirements to address most of the concerns raised in NUREG-1449.

#	Topic	Company	Comment	Response
44	Inspection	ENTERGY OP. INC	Team inspections would stress and dilute industry resources aimed at safe outage operations.	This is a valid concern. However, the staff pursues a policy of minimizing impact on licensees in the management of all its inspection activities.
45	Inspection	NUMARC	Team inspections have too adverse an impact, e.g., stress during outage. TI 2515/113 inspection is reasonable	The staff is considering this approach in its current regulatory analysis, as noted in Section 8.2.
46	Inspection	PACIFIC G&E	NRC should inspect each utility's implementation of NUMARC 91-06 and INPO initiatives.	The staff is considering the need for this action as part of its current regulatory analysis.
47	Instrumentation	GEORGIA POWER	A system like SPDS for shutdown would be costly and has little safety benefit.	The staff is currently considering the need for improved instrumentation in its current regulatory analysis.
48	Instrumentation	NUMARC	Broadening GL 88-17 requirements for instrumentation is not justified.	The staff is currently considering whether or not broadening GL 88-17 requirements for instrumentation is warranted.
49	Analysis	BNL	Revise NUREG-1449, Section 6.8 (rapid boron dilution), based on final version of NUREG/CR-5819.	Revisions have been incorporated.
50	Analysis	BWROG	PRA results in Chapter 4 are based on obsolete information. Use Surry/Grand Gulf results when available.	The staff recognizes the ages of these studies and has viewed the results of the studies accordingly. They are being retained in the report to document past work in this area.
51	Analysis	BWROG	Revise Figure 4.1 of NUREG-1449 to refer to boiling rather than core damage.	Footnote has been added to Figure 4.1 to clarify criterion for assuming core damage.
52	Analysis	BWROG	We disagree with statement that BWR Mark I/II secondary containments offer little protection.	The staff has reviewed the BWROG analysis. The results are reasonable given the assumptions (less conservative than staff). However, the staff continues to believe there are credible shutdown accidents that could lead to secondary containment failure.
53	Analysis	BWROG	The calculation showing BWR secondary containment failure in Section 6.9.1 is based on unrealistic assumptions.	See response to comment 52.
54	Analysis	DETROIT EDISON	We disagree with the staff finding that BWR Mark I/II secondary containments offer little protection.	See response to comment 52.
55	Analysis	ENTERGY OP. INC	The time to drain the Grand Gulf vessel from a flooded condition is 13 hours for rupture of a 4" RWCU system drain pipe.	This is a reasonable estimate for flooded conditions. The staff assumed initial water level was just below the top of the steam separators, i.e., a normal, non-flooded condition. This accounts for the much shorter drain time of 1 hour.

Table C.1 (cont.)

No.	Subject	Organization	Comment	Response
56	Analysis	ENTERGY OP. INC	Our calculations show 6–12 hours are available to restore RHR capability and prevent a significant offsite release at Grand Gulf.	The staff agrees that this amount of time may be available for some initial plant states.
57	Analysis	ENTERGY OP. INC	A LOCA could lead to BWR core damage in a short time; but liquid holdup in the secondary containment would filter the release of radioactive material.	Liquid holdup may act as a filter depending on the release point. However, the staff's analysis of a severe core damage accident during shutdown in a BWR/6 does not indicate that liquid holdup is a significant barrier to offsite or onsite releases.
58	Analysis	NORTHEAST UTIL	The secondary containment release scenarios are not credible and should be removed from NUREG–1449.	See response to comment 52.
59	Analysis	NUMARC	The freeze seal failure analysis is not credible if the refueling cavity is flooded.	The staff agrees that a significant amount of time would be available to mitigate a draindown event when the cavity is flooded. However, stopping a leak from the bottom of the vessel would be difficult.
60	Analysis	NUMARC	The thimble tube seal failure analysis is not credible. Conservative assumptions make the time to uncover the core too short.	Conservatism in the analysis is acknowledged in the report.
61	Analysis	NUMARC	The secondary containment release analysis is not credible.	See response to comment 52.
62	Analysis	BWROG	10 CFR 50.59 reviews should be performed for use of nozzle dams, main steamline plugs, and other temporary mechanical seals.	The staff agrees.
63	Operations	DETROIT EDISON	Changes to the license examiners handbook should be evaluated and validated by licensees prior to being issued.	Comments from industry and the general public have been considered in developing the latest revision to the license examiners handbook.
64	Operations	GEORGIA POWER	NUMARC 91–06 will not reduce hours worked but rather reschedule work to times of lower risk (Sec. 6.3).	As discussed in Section 6.3, overtime can be managed effectively with good planning. Depending on circumstances, it may be acceptable to shift work schedules according to a plan that considers risk as opposed to reducing hours worked.
65	Operations	NUMARC	Assess industry actions before imposing new requirements in the areas of operations, training, and procedures.	The staff is considering this approach in its current regulatory analysis.

#	Topic	Commenter	Comment	Response
66	Operations	NUMARC	The staff should reassess current requirements contributing to the unavailability of ac power (e.g., diesel generator maintenance).	The staff agrees that there are tradeoffs between doing diesel generator maintenance during power operation versus shutdown. These issues would normally be examined as part of the implementation of the maintenance rule.
67	Operations	TVA	PRA should be the basis for picking scenarios for simulator training.	PRA can serve as a basis for developing scenarios for simulator training. However, considering the complex human element in shutdown events and the simplicity of PRA models for shutdown, operating experience may be a better basis for shutdown.
68	Operations	WESTING-HOUSE	Reconsider the position in NUREG-1449 regarding venting the RCS by lifting the vessel head on its studs.	Based on further technical review on this issue, Section 3.3.3 of the NUREG report has been revised.
69	Operations	BWROG	We recommend no regulation of outage planning and control until NUMARC 91-06 is implemented and assessed.	The staff is considering this approach in its current regulatory analysis.
70	Operations	ENTERGY OP. INC	The conclusion in Section 6.7.1.4 regarding 50.59 reviews for freeze seals is no longer valid. Remove it.	The staff recognizes that improvements have been made in this area, and that Mississippi Power and Light has developed a strong program. The conclusion in Section 6.7.1.4 pertaining to "50.59" reviews has been revised to reflect these improvements.
71	Operations	ENTERGY OP. INC	A full core offload would require 1,000 more fuel movements, adding 12 days to a typical Grand Gulf outage.	The staff has considered this information in its current regulatory analysis as necessary.
72	Operations	ENTERGY OP. INC	The safety benefit of fully offloading the core is not demonstrated because loss of fuel pool cooling is not addressed in NUREG-1449.	This is true. The studies in NUREG-1449 deal only with operations when there is fuel in the reactor vessel. The staff is not considering a specific requirement for full core offload in its current regulatory analysis.
73	Outage Planning	ENTERGY OP. INC	New requirements are not appropriate now. Give industry efforts time to take effect.	The staff is considering this approach in its current regulatory analysis.
74	Outage Planning	NUMARC	Regulatory requirements on outage planning and control are unnecessary and duplicate industry efforts.	See response to comment 37.
75	Outage Planning	S. CAROLINA ELC	Regulatory requirements for outage planning and control are unnecessary and duplicate industry efforts.	See response to comment 37.
76	Regulating Action	TVA	Rulemaking is not necessary because shutdown operations are covered under 10 CFR 50 (Appendix B).	See response to comment 43.

Table C.1 (cont.)

No.	Subject	Organization	Comment	Response
77	Outage Planning	VEPCO	NUMARC 91-06 has been implemented at Surry and North Anna; NRC inspection (4/92) did not identify any problems.	The staff has observed that utilities have realized the positive effects of implementation of NUMARC 91-06 and has reflected this in the final report.
78	Outage Planning	WESTING-HOUSE	Hold requirements for outage planning and control in abeyance until NUMARC 91-06 is implemented and assessed.	The staff is considering this approach in its current regulatory analysis.
79	Tech Specs	BWROG	Current BWR/4 STS already provide TS improvements proposed by the staff for RHR and ECCS in BWRs.	The staff is considering this in its assessment of the need to change standard technical specifications that apply to nonpower operations.
80	Tech Specs	ENTERGY OP. INC	The technical basis for a requirement to close BWR/Mark III containment is not provided in NUREG-1449.	This is true. The staff has to show that such a requirement provides a cost-justified safety enhancement prior to imposition. This was not done in NUREG-1449.
81	Tech Specs	ENTERGY OP. INC	Current procedures for containment closure in PWRs are adequate. Requiring closure before boiling is too restrictive.	The staff is currently evaluating whether a technical specification governing containment integrity in PWRs during cold shutdown and refueling is warranted.
82	Tech Specs	ENTERGY OP. INC	The staff should coordinate development of new TS with appropriate industry groups.	Comments from industry and the general public on any proposed changes to standard technical specifications will be solicited and considered.
83	Tech Specs	ILLINOIS POWER	The containment closure requirement being considered for BWR/Mark III designs should be deleted due to its high cost/benefit ratio.	See response to comment 80.
84	Tech Specs	NUMARC	The TS improvements being considered by the staff would complement industry initiatives.	The staff is currently evaluating whether changes to technical specifications to address shutdown issues are warranted.
85	Tech Specs	NUMARC	We agree that the containment sump should be available to mitigate a LOCA; but TS should be flexible.	The staff is considering specific requirements regarding availability of the containment sump in PWRs during shutdown in its current regulatory analysis.
86	Tech Specs	NUMARC	RHR TS should require that support systems be FUNCTIONAL not OPERABLE.	The staff is considering specific requirements regarding the availability of systems that support the residual heat removal system in its current regulatory analysis. The staff is considering the desire for operational flexibility.

No.	Category	Organization	Comment	Response
87	Tech Specs	NUMARC	Get NSSS input from NSSS Owners Groups when developing new TS.	The staff met with each NSSS owners group individually in June 1992 to discuss shutdown issues. Staff expects to meet again with NSSS owners groups should new proposed requirements be issued for comment by the industry and the general public.
88	Tech Specs	NUMARC	New TS for ac power should be flexible and recognize need to perform maintenance. A requirement to have a minimum of 3 ac sources OPERABLE is acceptable.	The staff is considering specific requirements regarding ac power in its current regulatory analysis of potential requirements.
89	Tech Specs	NUMARC	The staff should not require 2 ECCS trains to be OPERABLE in reduced inventory because it is not justified and conflicts with requirements for LTOP systems.	The staff is considering the need for redundant trains of ECCS and the potential conflict with LTOP requirements in its current regulatory analysis of potential requirements.
90	Tech Specs	NUMARC	Only require PWR containments to be closed in reduced inventory, not full containment integrity.	The staff is considering need for containment integrity versus "containment closure" in its current regulatory analysis.
91	Tech Specs	PACIFIC G&E	NRC should adopt TS proposed in NUREG-1449 with NUMARC clarifications.	The staff will solicit comments from the industry and the general public on new technical specifications if they are to be proposed.
92	Tech Specs	S. CAROLINA ELC	Containment closure for PWRs should be defined as in GL 88-17.	See response to comment 90.
93	Tech Specs	S. CAROLINA ELC	The LCO for RHR in reduced inventory should be 1 train OPERABLE and 1 train FUNCTIONAL.	The staff is considering requirements for FUNCTIONAL versus OPERABLE systems in its current regulatory analysis.
94	Tech Specs	TVA	The staff should work with industry to develop new TS so as to achieve optimal safety and minimal impact.	See responses to comments 87 and 91.
95	Tech Specs	TVA	Relaxation of requirements to automatically go to cold shutdown when an RHR train is inoperable should be applied to BWRs as well as PWRs.	The staff will consider this in its current regulatory analysis.
96	Tech Specs	TVA	Browns Ferry does not have two trains of RHR which can be assigned to shutdown cooling at same time.	This comment refers to technical specifications which require two redundant trains of RHR. Given the Browns Ferry design, customized technical specifications for RHR may be most appropriate.
97	Tech Specs	VEPCO	New TS for shutdown conditions are warranted. Industry should participate fully in the development of new TS.	See responses to comments 87 and 91.

Table C.1 (cont.)

No.	Subject	Organization	Comment	Response
98	Tech Specs	WESTING-HOUSE	In developing new TS, the staff should seek NSSS Owners Group input through NUMARC.	The staff agrees. See responses to comments 87 and 91.
99	Tech Specs	WESTING-HOUSE	New TS should be based on NRC shutdown PRAs and staff review of operating experience.	The staff is considering results of PRAs and operating experience along with traditional regulatory criteria of redundancy and diversity in evaluating potential changes to technical specifications.
100	Tech Specs	WESTING-HOUSE	New shutdown TS should be consistent with interim policy statement on TS and not based on the current STS.	The staff agrees. Any new proposal for changes to standard technical specifications will be consistent with the policy statement.

APPENDIX D

Abbreviations

ABWR	advanced boiling-water reactor	IIT	incident investigation team
ACRS	Advisory Committee on Reactor Safeguards	ILRT	integrated leak rate test
		INEL	Idaho National Engineering Laboratory
AEOD	Office for Analysis and Evaluation of Operational Data	INPO	Institute of Nuclear Power Operations
		IPE	individual plant examination
AFW	auxiliary feedwater	IRM	intermediate range monitor
AIT	augmented inspection team	ISLOCA	intersystem loss-of-coolant accident
ALARA	as low as reasonably achievable	K/A	knowledge and abilities
ALWR	advanced light-water reactor		
ANS	American Nuclear Society	LCO	limiting condition for operation
ANSI	American National Standards Institute	LER	licensee event report
APRM	average power range monitor	LOCA	loss-of-coolant accident
ASME	American Society of Mechanical Engineers	LOOP	loss of offsite power
ASP	accident sequence precursor	LP	low power
ATWS	anticipated transient without scram	LPCI	low-pressure coolant injection
		LPS	low-power/shutdown
BNL	Brookhaven National Laboratory	LPSI	low-pressure safety injection
B&W	Babcock and Wilcox	LTOP	low-temperature overpressure protection
BWR	boiling-water reactor		
		MC	manual chapter
CDF	core-damage frequency	MOV	motor-operated valve
CE	Combustion Engineering	MPC	maximum permissible concentration
CFR	Code of Federal Regulations		
CNRA	Committee on Nuclear Regulatory Activities	NEA	Nuclear Energy Agency
		NPRDS	nuclear plant reliability data system
CR	control rooms	NRC	Nuclear Regulatory Commission
CRD	control rod drive	NRR	Office of Nuclear Reactor Regulation
CRGR	Committee To Review Generic Requirement	NSAC	Nuclear Safety Analysis Center
		NSSS	nuclear steam supply system
CS	core spray	NUMARC	Nuclear Management and Resources Council
CST	condensate storage tank		
DG	diesel generator	OECD	Organization for Economic Cooperation and Development
DHR	decay heat removal		
		OGC	Office of the General Counsel
EAL	emergency action level	ORNL	Oak Ridge National Laboratory
ECC	emergency core cooling		
ECCS	emergency core cooling system	PORV	power-operated relief valve
EDG	emergency diesel generator	POS	plant operational state
EOP	Emergency Operating Procedures	PRA	probabilistic risk assessment
EPRI	Electric Power Research Institute	PWR	pressurized-water reactor
ESF	engineered safety features		
		RCIC	reactor core isolation cooling
FSAR	final safety analysis report	RCP	reactor coolant pump
FY	fiscal year	RCS	reactor coolant system
		RES	Office of Nuclear Regulatory Research
GDC	general design criteria	RHR	residual heat removal
GE	General Electric	RHRSW	residual heat removal service water
GL	generic letter	RPS	reactor protection system
		RPV	reactor pressure vessel
HPI	high-pressure injection	RV	reactor vessel

RWSP	refueling water storage pool	SRP	Standard Review Plan (NUREG-0800)
RWST	refueling water storage tank	SRV	safety-relief valve
		STS	standard technical specifications
SAIC	Science Applications International Corporation	SW	service water
SBO	station blackout	TAF	top of active fuel
SD	shutdown	TI	temporary instruction
SDC	shutdown cooling	TS	technical specification(s)
SFP	spent fuel pool		
SG	steam generator	VCT	volume control tank
SI	safety injection		
SNL	Sandia National Laboratories	W̲	Westinghouse
SRM	source range monitor	WNP-2	Washington Nuclear Plant 2
SRO	senior reactor operator		

NRC FORM 335
(2-89)
NRCM 1102,
3201, 3202

U.S. NUCLEAR REGULATORY COMMISSION

BIBLIOGRAPHIC DATA SHEET

(See instructions on the reverse)

1. REPORT NUMBER
(Assigned by NRC, Add Vol.,
Supp., Rev., and Addendum Numbers, if any.)

NUREG-1449

2. TITLE AND SUBTITLE

Shutdown and Low-Power Operation at Nuclear Power Plants in the United States

Final Report

3. DATE REPORT PUBLISHED

MONTH	YEAR
September	1993

4. FIN OR GRANT NUMBER

5. AUTHOR(S)

6. TYPE OF REPORT

Technical

7. PERIOD COVERED (Inclusive Dates)

8. PERFORMING ORGANIZATION – NAME AND ADDRESS (If NRC, provide Division, Office or Region, U.S. Nuclear Regulatory Commission, and mailing address; if contractor, provide name and mailing address.)

Division of Systems Safety and Analysis
Office of Nuclear Reactor Regulation
U.S. Nuclear Regulatory Commission
Washington, DC 20555-0001

9. SPONSORING ORGANIZATION – NAME AND ADDRESS (If NRC, type "Same as above"; if contractor, provide NRC Division, Office or Region, U.S. Nuclear Regulatory Commission, and mailing address.)

Same as above

10. SUPPLEMENTARY NOTES

11. ABSTRACT (200 words or less)

The report contains the results of the NRC Staff's evaluation of shutdown and low-power operations at U.S. commercial nuclear power plants. The report describes studies conducted by the staff in the following areas: operating experience related to shutdown and low-power operations, probabilistic risk assessment of shutdown and low-power conditions and utility programs for planning and conducting activities during periods the plant is shut down. The report also documents evaluations of a number of technical issues regarding shutdown and low-power operations performed by the staff, including the principal findings and conclusions. Potential new regulatory requirements are discussed, as well as potential changes in NRC programs. A draft report was issued for comment in February 1992. This report is the final version and includes the responses to the comments along with the staff regulatory analysis of potential new requirements.

12. KEY WORDS/DESCRIPTORS (List words or phrases that will assist researchers in locating the report.)

Shutdown
Low-Power
Operations
Risk
Safety

13. AVAILABILITY STATEMENT

Unlimited

14. SECURITY CLASSIFICATION

(This Page)

Unclassified

(This Report)

Unclassified

15. NUMBER OF PAGES

16. PRICE

NRC FORM 335 (2-89)